Additive Manufacturing and 3D Printing Technology

Additive Manufacturing and 3D Printing Technology

Principles and Applications

G.K. Awari, C.S. Thorat, Vishwjeet Ambade,
and D.P. Kothari

CRC Press
Taylor & Francis Group
Boca Raton London New York

CRC Press is an imprint of the
Taylor & Francis Group, an **informa** business

CRC Press
Boca Raton and London
First edition published 2021

by CRC Press
6000 Broken Sound Parkway NW, Suite 300, Boca Raton, FL 33487-2742

and by CRC Press
2 Park Square, Milton Park, Abingdon, Oxon, OX14 4RN

© 2021 Taylor & Francis Group, LLC

CRC Press is an imprint of Taylor & Francis Group, LLC

The right of G.K. Awari, C.S. Thorat, Vishwjeet Ambade, and D.P. Kothari to be identified as the authors of the editorial material, and of the authors for their individual chapters, has been asserted in accordance with sections 77 and 78 of the Copyright, Designs and Patents Act 1988.
Reasonable efforts have been made to publish reliable data and information, but the author and publisher cannot assume responsibility for the validity of all materials or the consequences of their use. The authors and publishers have attempted to trace the copyright holders of all material reproduced in this publication and apologize to copyright holders if permission to publish in this form has not been obtained. If any copyright material has not been acknowledged please write and let us know so we may rectify in any future reprint.

Library of Congress Cataloging-in-Publication Data

Names: Awari, G.K., editor.
Title: Additive manufacturing and 3D printing technology : principles and
    applications / edited by G.K. Awari, C.S. Thorat, Vishwjeet Ambade, and
    D.P. Kothari.
Description: First edition. | Boca Raton, FL : CRC Press/Taylor & Francis
    Group, LLC, 2021. | Includes bibliographical references and index.
Identifiers: LCCN 2020038955 (print) | LCCN 2020038956 (ebook) | ISBN
    9780367436223 (hardback) | ISBN 9781003013853 (ebook)
Subjects: LCSH: Additive manufacturing. | Three-dimensional printing.
    Classification: LCC TS183.25 .A34 2021 (print) | LCC TS183.25 (ebook) |
    DDC 621.9/88--dc23
LC record available at https://lccn.loc.gov/2020038955
LC ebook record available at https://lccn.loc.gov/2020038956

ISBN: 978-0-367-43622-3 (hbk)
ISBN: 978-1-003-01385-3 (ebk)

Typeset in Times
by Deanta Global Publishing Services, Chennai, India

Instructors: Access the Support Material: routledge.com/9780367436223

This book is dedicated to the late Smt Shashikala K. Awari
for her heavenly blessings and encouragement.

Contents

Preface..xvii
Acknowledgments..xix
Authors..xxi
Glossary xxv

Chapter 1   Introduction to Additive Manufacturing and 3D Printing Technology 1

    1.1   Development of Additive Manufacturing 1
    1.2   Major Trends Shaping the Evaluation of 3D Printing 2
    1.3   Technology Improvement 3
    1.4   Active 3D Printing Market 4
    1.5   3D Printing Processes 5
          1.5.1   Material Extrusion..6
                1.5.1.1   Fused Deposition Modeling (FDM) 6
          1.5.2   Vat Photopolymerization...7
                1.5.2.1   Stereolithography (SLA)..............................8
                1.5.2.2   Digital Light Processing (DLP)...................8
          1.5.3   Powder Bed Fusion 8
                1.5.3.1   Powder Bed Fusion (Polymers)....................8
                1.5.3.2   Powder Bed Fusion (Metals)......................10
          1.5.4   Material Jetting 11
                1.5.4.1   Material Jetting (MJ)..................................12
                1.5.4.2   Drop on Demand (DOD) 12
          1.5.5   Binder Jetting 12
                1.5.5.1   Sand Binder Jetting....................................13
                1.5.5.2   Metal Binder Jetting 14
          1.5.6   Sheet Lamination 14
                1.5.6.1   Materials Used in Sheet Lamination 14
          1.5.7   Directed Energy Deposition...15
    1.6   Classification of Additive Manufacturing Systems 17
          1.6.1   Liquid-Based 17
          1.6.2   Solid-Based 17
          1.6.3   Powder-Based 18
    1.7   Advantages and Limitations...18
    1.8   Additive vs. Conventional Manufacturing Processes..............18
    1.9   Applications...19
    1.10 Exercises..19
    1.11 Multiple-Choice Questions..21

Chapter 2   CAD for Additive Manufacturing 25

    2.1   Introduction 25
    2.2   Preparation of CAD Models: The STL File 25
         2.2.1   STL File Format, Binary/ASCII............................26
              2.2.1.1   Format Specifications 28
              2.2.1.2   STL ASCII Format....................................29
              2.2.1.3   STL Binary Format 30
         2.2.2   Creating STL Files from a CAD System 30
         2.2.3   Calculation of Each Slice Profile 32
         2.2.4   Technology-Specific Elements 37
    2.3   Problems with STL Files..39
    2.4   STL File Manipulation 41
         2.4.1   Viewers..42
         2.4.2   STL Manipulation on the AM Machine 43
    2.5   Beyond the STL File..44
         2.5.1   Direct Slicing of the CAD Model 44
         2.5.2   Color Models..45
         2.5.3   Multiple Materials 45
         2.5.4   Use of STL for Machining 45
    2.6   Additional Software to Assist AM 46
         2.6.1   Survey of Software Functions 46
         2.6.2   AM Process Simulations Using Finite Element
              Analysis..47
    2.7   The Additive Manufacturing File Format 49
    2.8   Solved Example 50
    2.9   Exercises...51
    2.10 Multiple-Choice Questions..52

Chapter 3   Liquid-Based Additive Manufacturing Systems 55

    3.1   3D Systems Stereolithography Apparatus (SLA)....................55
    3.2   Stratasys PolyJet 59
         3.2.1   Fused Deposition Modeling from Stratasys 60
         3.2.2   Material Jetting Machines..61
    3.3   3D Systems' Multi-Jet Printing System (MJP) 62
    3.4   EnvisionTEC's Perfactory 63
    3.5   CMET's Solid Object Ultraviolet-Laser Printer (SOUP) 65
    3.6   EnvisionTEC's Bioplotter 66
    3.7   RegenHU's 3D Bioprinting 67
    3.8   Rapid Freeze Prototyping...69
    3.9   FDM 3D Printing for Zygomatic Implant Placement
         Mock Surgery for Prosthodontic Dentistry: A Case Study 70
         3.9.1   Zygomatic Implant 70
         3.9.2   Conclusion..72

3.10   Exercises 72
3.11   Multiple-Choice Questions.. 72

Chapter 4   Solid-Based Additive Manufacturing Systems 75

4.1   Stratasys Fused Deposition Modeling (FDM) 75
4.2   Solidscape's BenchTop System.. 75
4.3   Mcor Technologies' Selective Deposition Lamination (SDL)...... 77
4.4   Cubic Technologies' Laminated Object Manufacturing
       (LOM) 80
       4.4.1   Working Process.. 82
       4.4.2   Applications... 83
4.5   Ultrasonic Consolidation... 83
4.6   Exercises... 86
4.7   Multiple-Choice Questions.. 86

Chapter 5   Powder-Based Additive Manufacturing Systems............................. 89

5.1   3D Systems' Selective Laser Sintering (SLS) 89
       5.1.1   Technology 89
       5.1.2   Materials.. 90
       5.1.3   Powder Production 91
       5.1.4   Sintering Mechanisms 91
       5.1.5   Advantage.. 91
       5.1.6   Application 92
5.2   3D Systems' ColorJet Printing (CJP) Technology................ 92
       5.2.1   Technology 92
5.3   EOS's EOSINT System 93
       5.3.1   About EOS (Electro Optical Systems) 93
       5.3.2   EOSINT M 280 System 93
5.4   Optomec's Laser Engineered Net Shaping (LENS) and
       Aerosol Jet System.. 95
       5.4.1   About Optomec 95
       5.4.2   Laser Engineered Net Shaping (LENS) 95
       5.4.3   How the LENS System Works 95
5.5   Arcam's Electron Beam Melting (EBM)............................... 95
5.6   Concept Laser's LaserCUSING... 97
       5.6.1   Concept Laser's Patented LaserCUSING
                3D-Printing Technology... 97
       5.6.2   Working of LaserCUSING 3D-Printing Technology 98
       5.6.3   Additional Features of LaserCUSING
                3D-Printing Technology... 100
       5.6.4   Advantages of LaserCUSING 3D-Printing
                Technology 100
5.7   SLM Solutions' Selective Laser Melting (SLM).................. 100

5.8   Exercises... 102
5.9   Multiple-Choice Questions... 102

Chapter 6   Materials in Additive Manufacturing... 107
6.1   Choosing Materials for Manufacturing............................... 107
        6.1.1   Application... 107
        6.1.2   Asthetics... 108
        6.1.3   Function.. 108
        6.1.4   Certifications... 108
                6.1.4.1   Nylon.. 108
                6.1.4.2   ABS (Acrylonitrile Butadiene Styrene).... 109
                6.1.4.3   Resin... 109
                6.1.4.4   PLA (Polylactic Acid)............................ 110
                6.1.4.5   Gold and Silver 110
                6.1.4.6   Stainless Steel..................................... 111
                6.1.4.7   Titanium.. 111
                6.1.4.8   Ceramics... 111
                6.1.4.9   PET/PETG.. 112
                6.1.4.10   HIPS (High Impact Polystyrene).............. 112
                6.1.4.11   Thermoplastics..................................... 113
                6.1.4.12   Thermosets (Resins) 114
                6.1.4.13   Metals 114
6.2   Multiple Materials 114
        6.2.1   Multi-Material and Composite Additive
                Manufacturing Methods 115
                6.2.1.1   Stereolithography Methods..................... 115
                6.2.1.2   Binder Jetting Methods......................... 115
                6.2.1.3   Extrusion-Based Printing Methods 116
                6.2.1.4   Material Jetting Printing Methods 118
6.3   Metal AM Processes and Materials 119
        6.3.1   Powder-Bed Systems 120
        6.3.2   Powder-Fed Systems.. 120
        6.3.3   Direct Metal Laser Sintering (DMLS) Materials 121
        6.3.4   Selective Laser Sintering (SLS) Materials 123
        6.3.5   Stereolithography (SLA) Materials 124
6.4   Composite Materials... 127
        6.4.1   Generic Composite Materials................................. 128
        6.4.2   Composite Materials for 3D Printing Processes 129
6.5   Biomaterials, Hierarchical Materials, and Biomimetics 129
        6.5.1   Biomaterials 129
                6.5.1.1   Biomedical... 130
        6.5.2   Hierarchical Materials and Biomimetics 130
        6.5.3   Biomimetics... 133
                6.5.3.1   Definition of Biomimetic Material 135
                6.5.3.2   History of Biomimetic Materials............. 135

6.6   Ceramics and Bio-Ceramics.. 137
       6.6.1   Ceramics.. 137
       6.6.2   Bio-Ceramics.. 138
6.7   Shape-Memory Materials, 4D Printing, and Bio-Active
       Materials.. 139
       6.7.1   Shape-Memory Materials...................................... 139
               6.7.1.1   How Does Shape-Memory Work............. 140
               6.7.1.2   Shape-Memory Alloys........................... 141
               6.7.1.3   Shape-Memory Polymers 142
               6.7.1.4   Shape-Memory Composites.................... 146
               6.7.1.5   Shape-Memory Hybrids......................... 146
       6.7.2   4D Printing and Bio-Active Materials 150
6.8   Advanced AM Materials 151
6.9   Support Materials 152
       6.9.1   Build Material Supports... 153
               6.9.1.1   Materials as Their Own Support 153
       6.9.2   Quick Removal: Breakaway Supports...................... 154
       6.9.3   Best Quality: Soluble Supports 155
       6.9.4   Supporting with the Original Material.................... 156
       6.9.5   Supporting with PVA Filament 156
       6.9.6   Supporting with PVA+ Filament 156
6.10   Exercises... 156
6.11   Multiple-Choice Questions.. 157

Chapter 7   Applications and Examples 159
7.1   Application–Material Relationship 159
       7.1.1   Polymer.. 159
       7.1.2   Metal.. 161
7.2   Finishing Processes 162
       7.2.1   Plating.. 163
       7.2.2   Sanding.. 163
       7.2.3   Bead Blasting 164
       7.2.4   Shot Peening.. 164
       7.2.5   Heat Treatments 164
       7.2.6   Vibratory Systems 165
       7.2.7   Tumbling 165
       7.2.8   Vapor Smoothing.. 166
       7.2.9   Solvent Dipping.. 166
       7.2.10 Epoxy Coating.. 166
       7.2.11 Epoxy Infiltration 166
       7.2.12 Painting.. 167
7.3   Applications in Design 167
       7.3.1   CAD Model Verification 167
       7.3.2   Visualizing Objects... 167
       7.3.3   Proof of Concept 167

7.4   Applications in Engineering, Analysis, and Planning........... 168
        7.4.1   Scaling 168
        7.4.2   Form and Fit.. 168
        7.4.3   Flow Analysis.. 168
        7.4.4   Pre-Production Parts 168
7.5   Applications in Manufacturing and Tooling 168
        7.5.1   Direct Soft Tooling 168
        7.5.2   Indirect Soft Tooling 168
        7.5.3   Direct Hard Tooling 169
        7.5.4   Indirect Hard Tooling... 169
7.6   Aerospace Industry.. 169
        7.6.1   The Benefits of 3D Printing for Aerospace and
                 Defense 170
                 7.6.1.1   Low-Volume Production 170
                 7.6.1.2   Weight Reduction...................................... 170
                 7.6.1.3   Material Efficiency 170
                 7.6.1.4   Consolidation of the Part.......................... 171
                 7.6.1.5   Maintenance and Repair............................ 171
7.7   Automotive Industry.. 171
        7.7.1   The Benefits of 3D Printing for Automotive 172
                 7.7.1.1   Faster Product Development...................... 172
                 7.7.1.2   Greater Design Flexibility 172
                 7.7.1.3   Customization.. 172
                 7.7.1.4   Create Complex Geometries..................... 172
7.8   Jewelry Industry 172
        7.8.1   Investment Casting 173
        7.8.2   Direct Printing.. 176
7.9   Coin Industry... 178
7.10   Tableware Industry 180
7.11   Geographic Information System (GIS) Applications 180
7.12   Arts and Architecture... 182
        7.12.1   Benefits of 3D Printing for Architects 183
7.13   Construction 183
7.14   Fashion and Textiles 184
7.15   Weapons 186
7.16   Musical Instruments 186
7.17   Food... 187
7.18   Movies 188
7.19   Design and Development of a Prosthetic Hand through
        3D Printing: Case Study.. 189
        7.19.1   Conclusion.. 194
                 7.19.1.1   Prototype Assessment............................. 194
7.20   Exercises.. 195
7.21   Multiple-Choice Questions.. 196

Chapter 8   Additive Manufacturing Equipment 199

   8.1   Process Equipment—Design and Process Parameters.......... 199
         8.1.1   Seven Distinct AM Processes 199
                 8.1.1.1   Powder Bed Fusion 199
                 8.1.1.2   Directed Energy Deposition 200
                 8.1.1.3   Binder Jetting.. 200
                 8.1.1.4   Sheet Lamination 200
                 8.1.1.5   Material Extrusion....................................... 200
                 8.1.1.6   Material Jetting.. 201
                 8.1.1.7   Vat Photopolymerization 201
         8.1.2   Designing for 3D Printing 201
   8.2   Governing Bonding Mechanism 201
         8.2.1   Overview of the Bonding Process 202
         8.2.2   Bond Mechanisms 204
   8.3   Common Faults and Troubleshooting 204
         8.3.1   The Printer Is Working but Nothing Is Printing 205
                 8.3.1.1   The Problem—Out of Filament................ 205
                 8.3.1.2   The Cause 205
                 8.3.1.3   The Solution.. 205
         8.3.2   Nozzle Is Too Close to the Print Bed 205
                 8.3.2.1   The Problem.. 205
                 8.3.2.2   The Cause 206
                 8.3.2.3   The Solution.. 206
         8.3.3   Over-Extrusion 206
                 8.3.3.1   The Problem—Print Looks Droopy
                           and Stringy... 206
                 8.3.3.2   The Cause 207
                 8.3.3.3   The Solution.. 207
         8.3.4   Incomplete and Messy Infill..................................... 207
                 8.3.4.1   The Problem.. 207
                 8.3.4.2   The Cause 208
                 8.3.4.3   The Solution.. 208
         8.3.5   Warping... 208
                 8.3.5.1   The Problem—Bending............................. 208
                 8.3.5.2   The Cause 208
                 8.3.5.3   The Solution.. 208
         8.3.6   Messy First Layer 209
                 8.3.6.1   The Problem.. 209
                 8.3.6.2   The Cause 209
                 8.3.6.3   The Solution.. 210
         8.3.7   Elephant's Foot... 210
                 8.3.7.1   The Problem.. 210
                 8.3.7.2   The Cause 210
                 8.3.7.3   The Solution.. 210

8.3.8   Print Looks Deformed and Melted 211
          8.3.8.1   The Problem.. 211
          8.3.8.2   The Cause 211
          8.3.8.3   The Solution.. 211
8.3.9   Snapped Filament.. 211
          8.3.9.1   The Problem.. 211
          8.3.9.2   The Cause 212
          8.3.9.3   The Solution.. 212
8.3.10   Getting Cracks in Tall Objects............................. 213
          8.3.10.1   The Problem... 213
          8.3.10.2   The Cause 213
          8.3.10.3   The Solution... 213
8.4   Process Design 213
        8.4.1   Creation End Evaluation of Support Structure......... 214
        8.4.2   Additive Manufacturing Preparation 214
        8.4.3   Validation of Build Time and Cost 215
        8.4.4   Additive Manufacturing Simulation....................... 215
8.5   Low Cost, Rapid Deployment Wireless Patient
        Monitoring System Developed with Additive
        Manufacturing Equipment: Case Study 216
        8.5.1   Conclusion 219
8.6   Exercises.. 219
8.7   Multiple-Choice Questions.. 220

Chapter 9   Post-Processing 221

9.1   Introduction 221
9.2   Support Material Removal 221
        9.2.1   Natural Support Post-Processing............................ 222
        9.2.2   Synthetic Support Removal.................................... 223
                  9.2.2.1   Supports Made from the Build Material..... 224
                  9.2.2.2   Supports Made from Secondary
                              Materials... 225
9.3   Surface Texture Improvements.. 226
9.4   Accuracy Improvements... 226
        9.4.1   Sources of Inaccuracy 226
        9.4.2   Model Pre-Processing to Compensate for
                  Inaccuracy 227
        9.4.3   Machining Strategy 228
                  9.4.3.1   Adaptive Raster Milling 228
                  9.4.3.2   Sharp Edge Contour Machining.............. 230
                  9.4.3.3   Hole Drilling... 232
9.5   Esthetic Improvements 232
9.6   Preparation for Use as a Pattern 233
        9.6.1   Investment Casting Patterns................................... 233

	9.6.2	Sand Casting Patterns	234
	9.6.3	Other Pattern Replication Methods	235
9.7	Property Enhancements Using Non-Thermal Techniques		236
9.8	Property Enhancements Using Thermal Techniques		237
9.9	Exercises		240
9.10	Multiple-Choice Questions		240

Chapter 10 Product Quality 243

10.1	Building the Part	243
	10.1.1 Powder Reclamation	243
10.2	Post-Processing and Finishing	244
10.3	Bulk Deposit Defects	247
10.4	Dimensional Accuracy, Shrinkage, and Distortion	254
10.5	Inspection, Quality, and Testing of AM Metal Parts	255
	10.5.1 Nondestructive Test Methods	255
	10.5.2 Destructive Test Methods	258
	10.5.3 Form, Fit, Function, and Proof Testing	261
10.6	Standards and Certification	262
10.7	Key Takeaway Points	264
10.8	Overview of 4D Printing	264
	10.8.1 A Definition of 4D Printing	264
	10.8.1.1 The Difference between 4D and 3D Printing	265
	10.8.2 Potential Applications for 4D Printing	265
10.9	Exercises	265
10.10	Multiple-Choice Questions	266

Interview Preparedness 267

References 275

Index 279

Preface

Additive manufacturing (AM) and 3D-printing technology are growing quickly and in multiple directions. The level of activity is at an all-time high as researchers, investors, company management, and government agencies try to predict where it is headed. Many believe it is the next "big thing." Organizations of all types and sizes are trying to understand the role they might play. Companies are at work utilizing machines and materials for the direct manufacture of parts that go into final products while assembling. Aerospace companies are qualifying additive manufacturing processes and materials and certifying new designs for flight.

The advantages of AM are in the design and redesign of the parts. It is possible to consolidate many individual parts of an assembly (as many as 15 or more) into a single, complex part. Such an approach to design eliminates part numbers, inventory, assembly, labor, and inspection. It is possible to redesign parts with relatively thin skins that include internal lattice/mesh structures instead of solid material throughout, which can substantially reduce the amount of material, weight, and build time. In some cases, the amount of material and weight has been reduced by more than 50% using these techniques.

The challenges ahead are system reliability and process repeatability, especially when using AM for manufacturing. System manufacturers are addressing these challenges with real-time process monitoring and control software, but much work is yet to be done in the field. The current limitations in build speed and maximum part size are challenges, too. Manufacturers are developing systems with larger build volumes and methods that increase throughput. The aim of compiling this book has been to give a working knowledge to all engineering students of the important details of additive manufacturing and 3D printing, technology, and materials used in manufacturing, and several other major topics in a systematic way. The book is written in a clear and easy-to-read style, presenting fundamentals of additive manufacturing at a level that can be quickly grasped by a beginner. The book addresses these challenges and provides readers at all levels with an insight into additive manufacturing and 3D-printing technology.

The comprehensive subject matter is organized into ten chapters: 1. Introduction to Additive Manufacturing and 3D Printing Technology; 2. CAD for Additive Manufacturing; 3. Liquid-Based Additive Manufacturing Systems; 4. Solid-Based Additive Manufacturing Systems; 5. Powder-Based Additive Manufacturing Systems; 6. Materials in Additive Manufacturing; 7. Applications and Examples; 8. Additive Manufacturing Equipment; 9. Post-Processing; 10. Product Quality; followed by a brief guide to interview preparedness.

The book covers several applications of 3D printing, including typical examples, from an examination point of view, and also industrial case studies. The text uniquely addresses complete solutions to additive manufacturing without ignoring any relevant topics. The book is mainly aimed at engineering courses at the diploma, graduate, and postgraduate levels offered in most universities in India and USA.

This book will serve as a major resource for students of Mechanical Engineering, Production Engineering, Design Engineering, Electrical Engineering, Electronics Engineering, and Industrial Engineering.

Multiple-choice questions and exercises have been added to the end of each chapter to make the book a comprehensive unit in all respects. This book is also useful to prepare students for competitive examinations such as GATE, IES, UPSC, and other public sector undertakings.

The main characteristics of this book are

i) The subject of additive manufacturing and 3D printing is itself in the innovation phase and it is assumed that the reader's goal is to achieve a suitable balance of cost, schedule, and quality.

ii) The book has been written in luculent language, which can be understood by students at the diploma level (junior level) of engineering courses.

iii) The book presents a selection of various additive manufacturing techniques suitable for near term application, with sufficient technical background to understand the domain of applicability and to consider variations to suit technical and organizational constraints.

iv) The new innovative 3D-printing systems and composite materials developed in the past decade are incorporated into the book and the future view of 4D printing is presented to the readers.

v) The book promotes a vision of additive manufacturing and applications as integral to modern manufacturing engineering practices, equally as important and technically demanding as other aspects of development. This vision is consistent with current thinking on the subject. Several applications of 3D printing have been added to the book.

vi) The book consists of case studies, self-explanatory figures, and photographs of prototypes developed in the laboratory; it is inspiring for budding manufacturing entrepreneurs.

Acknowledgments

The achievement of a mission is never a solo effort; it is the product of the important involvement of a variety of individuals in direct or indirect ways who have enabled us to make it a success. We would like to extend our appreciation and recognize the guiding lights who have helped us to accomplish this mission. *Additive Manufacturing and 3D Printing Technology: Principles and Applications* is the culmination of the authors' classroom and laboratory experiences.

We are grateful to Dr. Abhay Wagh, Director of Technical Education (DTE), Mumbai, (MS), India; Dr. Vinod Mohitkar, Director of MSBTE, Mumbai, India; Dr. Ram Nibudey, Joint Director of Technical Education, Regional Office Nagpur, India; Dr. M. B. Daigavane, Principal of Government Polytechnic, Nagpur, India; Prof. Deepak S. Kulkarni, from Government Polytechnic, Nagpur and Dr. S. W. Rajurkar, HOD (Mechanical Engineering) at Government Engineering College Chandrapur, India, for their constant inspiration and encouragement to develop the learning resources.

We are indebted to Dr. Mohan Gaikwad-Patil, Chairman, Gaikwad-Patil Group of Institutions, Nagpur (MS), India; Prof. Sandeep Gaikwad, Treasurer, Gaikwad-Patil Group of Institutions, Nagpur (MS), India; Mr. Mukul Pande, Director, Gaikwad-Patil Group of Institutions, Nagpur, India, for extending the laboratory facilities of the institute of Tulsiramji Gaikwad-Patil College of Engineering and Technology, Nagpur, India, to complete this project. We are thankful to Mr. Sumeet Gattewar, Director, Pye Technologies India; Dr. Priya Gupta; Dr. Saee Deshpande from VSPM Dental College, Nagpur, India; and Mr. Abhijeet Raut, Research Fellow, VNIT, Nagpur (MS), India, for their support in developing the case studies. We are also thankful to Mr. Vidyadhar Kshirsagar, Mr. Yogesh Ramteke, and Mr. Niteen Kakde from the Tulsiramji Gaikwad-Patil College of Engineering and Technology, Nagpur, India, for assisting us in developing the figures with CAD software. We have gained greatly in preparing the manuscript of this book by referring to several articles, journals, online sources, and open source material. We express our gratitude to all such authors, publications, and publishers, many of whom have been included in the bibliography. If someone has been left out unintentionally, we will seek their forgiveness.

The authors are very grateful to Prof. Dr. Jaji Varghese, Aryabhat Polytechnic, New Delhi, India; Prof. Dr. S. Velumani, Velarar College of Engineering, Erode (TN), India; Dr. Abhijeet Digalwar, Professor, BITS Pilani; Dr. D.K. Parbat, Government Polytechnic, Bramhapuri, India; Dr. N.V. Raut, Dr. S.R. Kukadapwar, Dr. D.N. Kongre, Prof. P.V. Rekhade, Dr. R.G. Chaudhari, Dr. G.V. Gotmare, Dr. K.S. Dixit, Prof. Ghormade, Prof. Pise, Prof. Dupare, Prof. R.B. Tirpude, Prof. V.S. Kumbhar from Government Polytechnic, Nagpur, India; and Dr. S.S. Chaudhari, Dr. Khedkar, from YCCE, Nagpur, India, for their consistent support and assistance in creating this book.

We sincerely thank our mentors, Dr. D.G. Wakde, Dr. L.B. Bhuyar, who have helped us and have been a source of inspiration. We are obliged to all leading

3D-printing manufacturers of this region for their backing and support. In preparing the manuscript of this book, we have benefited immensely from referring to many books, publications and online sources, such as website and open source material. We express our gratitude to all those authors, publications, and publishers; many of them have been listed in the bibliography. If anybody is left out inadvertently, we seek their pardon.

We thank CRC Press, Taylor & Francis Group, especially Ms. Cindy Renee Carelli and Ms. Erin Harris, who have kept our morale high, helped us in preparing and maintaining our schedules, facilitated the work, provided regular updates, and stood behind us patiently during this entire work. We also thank Ms. Jayanthi Chander for project managing the manuscript.

Indeed, we are thankful to our family members for their timely support in all the efforts of this book, without which it would not have seen the light of day. The efforts of our family member Mrs. Jaya Awari, Master Vedant Awari, Dr. Madhuri Thorat, Mrs. Nikita Ambade and Mrs. Shobha Kothari, are appreciated for helping us during the entire duration of this project. Last but most important, we bow our heads to the majesty of the Almighty God and our parents for making our experience one of the most technologically satisfying moments of our lives.

We hope that the book will serve the intent of its readers and that we will continue to receive their help. Suggestions to enhance the quality and style of the book are always welcome and accepted and integrated into future editions.

Authors

Dr. G.K. Awari earned a Bachelor of Engineering (BE) degree from RTM Nagpur University, Nagpur, Maharashtra, India, in 1991 and a Master of Engineering (ME) degree from Thapar University, Patiala, Punjab, India, in 1995, both in Mechanical Engineering. He completed his PhD at Sant Gadgebaba Amravati University, Amravati, Maharashtra, India, in 2007. He has more than 25 years of teaching experience at diploma, undergraduate, postgraduate, and research levels. He has taught various subjects such as Fluid Power and Machinery, Computer Graphics, Automation Engineering, Operation Research, Machine Design, and Automobile Engineering. His area of interest is graphical modeling of Computational Fluid Dynamics (CFD), Vehicle Dynamics.

He has 213 citations including total 33 international journal publications, 22 international conference publications, and 11 national conference publications, and three patents and one product developed for the industry to his name. Seventeen research scholars have completed PhDs in Mechanical Engineering under his supervision at three Indian universities. He is also a recipient of the "Best Principal Award" and "Best Paper Award" at various national and international conferences.

He has contributed to the development of academics as Board of Study (BOS) Member at Goa University, Goa, SG Amravati University, Amravati, India, and RTM Nagpur University, Nagpur, India. He is presently BOS Member in Yeshvantrao Chavan College of Engineering (YCCE, an autonomous institute), Nagpur, India, RTM Nagpur University, Nagpur, India, GH Raisoni University, Chhindwara (MP), India, and Chairman BOS of Automobile Engineering at Government Polytechnic (GP), Nagpur, India. He is also recognized by AICTE as "Margadarshak for NBA Accreditation" of mentee institutes.

He has authored eight books, including two books with CRC Press, Taylor & Francis Group, and others with renowned international publishers such as Mercury International Publication, New Delhi, and New Age India Publisher Ltd., New Delhi. He has also developed more than fifty video tutorials/e-content modules for the benefit of students/teachers and his style of presentation/lectures is appreciated by many staff and students.

Dr. Awari is eminent in the Co-Learning Process and Participative Management. He has been ranked first in merit position in MPSC throughout MS and currently he is Head of the Automobile Engineering Department at Government Polytechnic,

Nagpur, India. He is a rare combination of the best academician and administrator and is loved by students and staff members.

Dr. C.S. Thorat earned his BE degree from Government College of Engineering, Karad, Maharashtra, India, in 1985 and MTech degree from IIT Bombay, India, in 1986, both in Electrical Engineering. He completed his PhD at Sant Gadgebaba Amravati University, Amravati, Maharashtra, India, in 2005. He has more than 30 years of teaching experience at diploma, undergraduate, postgraduate, and research levels. He also has field experience at the Oil and Natural Gas Commission, India. He has guided five candidates through doctoral research work in Electrical Engineering and also developed the laboratory of 3D printing at the institute. He has more than 100 research publications and five patents to his name. He has successfully demonstrated the capabilities of 3D printing in additive manufacturing through product development. Dr. Thorat is currently Principal of Government Polytechnic, Jalna (MS), India.

Prof. Vishwjeet Ambade earned his M.Tech degree in Mechanical Engineering Design from Karamvir Dadasaheb Kannamwar College of Engineering (KDK), RTM Nagpur University, Nagpur, India. He graduated in Mechanical Engineering from Kavikulguru Institute of Technology and Science, Ramtek (KITS), Nagpur, India. He has more than ten years of experience in teaching different subjects within the discipline of Mechanical Engineering and presently he is working as Assistant Professor at the Mechanical Engineering department at Tulsiramji Gaikwad-Patil College of Engineering and Technology, Nagpur, India. He has guided many UG/PG projects and he is pursuing his PhD research work in 3D printing and its application at Gondwana University, India. He has contributed to more than ten international journal/conference papers. He has authored two books.

 **Dr. D.P. Kothari** is a Director Research and Professor, S B Jain Institute of Technology, Management and Research, Nagpur. Presently he is Honorary Adjunct Professor at VNIT, Nagpur, India. He earned his BE (Electrical) in 1967, ME (Power Systems) in 1969, and PhD in 1975 from Birla Institute of Technology and Sciences (BITS), Pilani, Rajasthan, India. From 1969 to 1977, he was involved in the teaching and development of several courses at BITS Pilani. Dr. Kothari served as Vice Chancellor, VIT, Vellore, Director in-charge, and Deputy Director (Administration) as well as Head in the Centre of Energy Studies at the Indian Institute of Technology, Delhi, India, and as Principal, Visvesvaraya Regional College of Engineering (VNIT), Nagpur, India. He was Visiting Professor at the Royal Melbourne Institute of Technology, Melbourne, Australia, for two years in 1982–1983 and 1989. He was also NSF Fellow at Perdue University, United States, in 1992.

Dr. Kothari, who is a recipient of the most Active Researcher Award, has published and presented 830 research papers in various national and international journals and conferences, guided 56 PhD scholars and 65 MTech students, and authored 67 books in Engineering and Technology with reputed publishers. He has delivered several keynote addresses and lectures at both national and international conferences. He has also delivered 42 video lectures on YouTube with around 40,000 hits!

Dr. Kothari is a Fellow of the National Academy of Engineering (FNAE), Fellow of the Indian National Academy of Science (FNASc), Fellow of the Institution of Engineers (FIE), Fellow of IEEE, and Hon. Fellow of ISTE. His many awards include the National Khosla Award for Lifetime Achievements in Engineering (2005) from IIT, Roorkee, India. The University Grants Commission (UGC), Government of India has bestowed upon him the UGC National Swami Pranavandana Saraswati Award (2005) in the field of education for his outstanding scholarly contributions.

Glossary

3D	3-Dimensional
3DP	3-DimensionalPrinting
A&D	Aerospace and Defense
ABS	Acrylonitrile Butadiene Styrene
AEC	Architects, Engineers, and Construction
AM	Additive Manufacturing
ANSI	American National Standards Institute
ASME	American Society of Mechanical Engineers
AWS	American Welding Society
BHGM	Biomimetic Hierarchical Graphene Material
BIM	Building Information Modeling
BJ	Binder Jetting
CAD	Computer-Aided Design
CAGR	Compound Annual Growth Rate
CFRP	Carbon Fiber Reinforced Plastic
CFRTP	Carbon Fiber Reinforced Thermoplastic
CHEM	Cold Hibernated Elastic Memory
CJP	ColorJet Printing
CLIP	Continuous Liquid Interphase Production
CMM	Coordinate Measurement Machines
CT	Computer Tomography
DED	Directed Energy Deposition
DIY	Do-It-Yourself
DLP	Digital Light Processing
DMD	Digital Micromirror Device/Direct Metal Deposition
DMLM	Direct Metal Laser Melting
DOD	Drop On Demand
DT	Destructive Test
EBF3	Electron Beam Free-Form Fabrication
EBM	Electron Beam Melting
EDM	Electrode Discharge Machining
EOS	Electro Optical Systems
ET	Eddy-Current Testing
EWI	Edison Welding Institute
FAA	Federal Aviation Administration
FDA	Food and Drug Administration
FDM	Fused Deposition Modeling
FEA	Finite Element Analysis
FFF	Fused Filament Fabrication
FGMs	Functionally Graded Materials
FRP	Fiber Reinforced Plastic

GFK	Glassfaserverstärkter Kunststoff
GFRP	Glass Fiber Reinforced Plastic
GIS	Geographic Information System
GRP	Glass Reinforced Plastic
HIP	Hot Isostatic Pressing
HIPS	High Impact Polystyrene
HP	Hewlett-Packard
IJP	Inkjet Printing
IPH	Indentation-Polishing-Heating
ISO	International Organization for Standardization
LDM	Low-Temperature Deposition
LENS	Laser Engineered Net Shaping
LMD	Laser Metal Deposition
LOM	Laminated Object Manufacturing
MBJ	Metal Binder Jetting
MDF	Medium-Density Fiberboard
MEMS	Micro-Electromechanical Systems
MJ	Material Jetting
MJP	Multijet Printing
MPVP	Mask Projection Vat Photopolymerization
MRI	Magnetic Resonance Imaging
MRO	Maintenance, Repair, and Overhaul
NDT	Nondestructive Test
NEMS	Nano-Electromechanical Systems
OSHA	Occupational Health and Safety Organization
PA	Polyamides
PAEK	Polyaryletherketons
PBD	Powder Bed Fusion
PC	Polycarbonate
PET	Polyethylene Terephthalate
PLA	Polylactic Acid
PPSF	Polyphenylsulfone
PS	Polystyrenes
PT	Penetrant Testing
PVDF	Polyvinylidene Fluoride
RFID	Radio Frequency Identification
RFP	Rapid Freeze Prototyping
RP	Rapid Prototyping
RT	Radiographic Testing
RTV	Room Temperature Vulcanization
SAE	Society of Automotive Engineers
SDL	Selective Deposition Lamination
SFF	Solid Freeform Manufacturing
SHS	Selective Heat Sintering
SL	Sheet Lamination

SLA	Stereolithography Apparatus
SLM	Selective Laser Melting
SLS	Selective Laser Sintering
SMAs	Shape-Memory Alloys
SMCs	Shape-Memory Composites
SMHs	Shape-Memory Hybrids
SMPs	Shape-Memory Polymers
SOUP	Solid Object Ultraviolet-Laser Printer
STL	Stereolithography
TiAl	Titanium Aluminum
TME	Temperature Memory Effect
TPE	Thermoplastic Elastomers
UAM	Ultrasonic Additive Manufacturing
UHMWPE	Ultra-High-Molecular-Weight Polyethylene
UT	Ultrasonic Testing
UV	Ultraviolet
VP	Vat Polymerization

1 Introduction to Additive Manufacturing and 3D Printing Technology

1.1 DEVELOPMENT OF ADDITIVE MANUFACTURING

Since the 1980s, 3D printing technology, also known as additive manufacturing, has existed in some form or another. Nevertheless, the technology was neither efficient nor cost-effective enough for most end-products or high-volume industrial manufacturing. Expectations are very high that these shortcomings are about to be eliminated. Additive manufacturing (AM) technology has emerged as a result of developments in a variety of technology sectors. As additive manufacturing continues to gain popularity and its technology rapidly evolves, designers are able to produce better goods faster and cheaper, without thinking about the limitations of conventional manufacturing processes. Unlike other industrial innovations, increased computing capacity and decreased mass storage costs paved the way for the processing of vast volumes of data typical of modern 3D computer-aided design (CAD) models within a realistic timeframe. Nowadays, researchers have become used to having powerful computers and other complex electronic devices around them, and often it can be difficult to understand how the pioneers struggled to build the first AM devices. This subject highlights some of the key moments in the growth of additive manufacturing technology, and how the various technologies converged to the point that they could be merged into AM machines will be explained. The AM technologies milestone will also be discussed. In addition, how the application of additive manufacturing has evolved to include greater functionality and a wider range of applications beyond the initial intention of prototyping will be discussed. Emphasis is also placed on how additive manufacturing affects the automotive and aerospace industries. The use of AM in product development is necessary to enable companies to compete with industry standards.

Additive manufacturing, better known as 3D printing (3DP) on the market, has evolved over the past 40 years. There is growing evidence that the advances in technology and materials have finally gone beyond the hype level. Thirty-six percent of businesses are either applying or planning to apply 3DP, according to a recent World Economic Year survey of various firms. Aerospace, defense, and automotive are the most specialized sectors to apply 3DP. ThreeD printing technology will advance through a loosely coordinated development in three areas: Printing and printing methods, design and printing software, and printing materials. However, awareness

of 3DP and willingness to leverage it for prototyping, tools, fixtures, and even finished products is increasing in other industries for a number of reasons.

- **Quality and speed**: As printer speed has increased, quality assurance tools
  embedded in printers enable better layer-by-layer validation of whether the
  printed product is within acceptable tolerances.
- **Availability of materials**: A wide range of materials and sources of materials are now available, creating more incentives for the industry to manufacture parts and goods. Many manufacturers and industries are now working
  with suppliers of materials to create their own material variations in order
  to meet their specific requirements or to improve quality.
- **Workforce knowledge**: The newest and youngest generation of designers
  and engineers is more knowledgeable about 3DP.
- **Product development**: 3DP improves time-to-market and shortens product
  design cycles.
- **Manufacturing**: 3DP reduces process time by using improved tools, a
  technology that tends to reduce waste.
- **Engineering and maintenance**: Maintenance processes are more flexible
  and may reduce maintenance costs.
- **Storage and warehousing**: Reduced inventory, logistic, and storage costs.
- **Aftermarket**: Improved flexibility in the supply of spare parts and
  decreased costs in the manufacture of spare parts.

1.2 MAJOR TRENDS SHAPING THE
EVALUATION OF 3D PRINTING

Several developments in the commercial and corporate sector have accelerated 3D
printing to the top in manufacturing industry. Approximately 25% of global companies uses 3D printing, with another 12% considering it, 3D printing technology is
one of the greatest inventions to date and is becoming a reality—with a large amount
of new opportunities and challenges.

The key business trends for the evolution of 3DP are as follows:

- **Individualization—customer co-creation**: There is significant growth
  and development in the modern economy for the modification and individualization of manufactured products according to consumer needs and
  requirements. Manufacturers responded by adding 3DP and choose to
  explore prospects for the use of 3DP in future applications.
- **Democratization—mass innovation and development**: 3DP makes it
  easier for individuals or collaborative teams to design or manufacture end-
  products and reduces barriers and challenges to innovation. The manufacturer of the product, which may be hampered in a traditional routine with
  difficult requisitioning processes and long logistical wait times, would now
  have the option of making the products faster. Design teams can "fax" their
  parts to work with substantial products for intra-company collaboration.

When the required engineering or design resources are not co-located or even virtually linked, companies have turned to crowdsourcing.

• **Sustainability—a circular economy**: There is a global movement toward sustainability both for the home and for corporations of all sizes. 3DP reduces transport costs when the 3D printer is placed close to the production line. There are also operating cost efficiencies when aircrafts are built with lighter materials. That said, as early adopters move toward the use of 3DP, there are often missing factors in the cost-benefit analysis: For example, power and heat are critical to 3DP processes; also, the advantage of easy iteration on designs that increase the amount of unrecyclable waste. As a result, the environmental benefits of 3DP do not come without care and planning, but can be achieved.

1.3 TECHNOLOGY IMPROVEMENT

The wide range of materials that can now be used is vast, including plastics, porcelain, ceramics, stainless steel, carbon, graphene, titanium, and many more metals. This list is not exhaustive—new material or alloy combinations are made every day. For parts considered suitable for 3DP by a company, there is a reduction in the supplier base, as the components no longer need to be sourced. There are four different patterns in 3DP technology.

• **Beyond prototyping applications, impacts process designs**: Previously, 3DP was initially limited to prototyping. Now, 3DP methods such as direct metal laser sintering, selective laser sintering, and electron beam melting have advanced 3DP in industrial applications and final assemblies. When considering the application of 3DP, the question that companies need to ask is: What network of supply chain assets and what mix of old and new processes will be optimal? Some processes may benefit from the input of a component made on demand with 3DP, but others may not be suitable for 3DP. One thing that 3DP is capable of delivering is more data that support the alignment of existing and new processes. When 3D printing is used to manufacture parts, that part of the process becomes digital, with each element of each part produced under continuous software control.

• **Lightweight materials**: Honeycombing is another 3DP method that allows especially lightweight parts to be printed. It is possible to build hollow parts or sections with an inner chamber that are connected in a way similar to the inside of a beehive. If companies or individuals outsource any 3D prints, one big question will be whether the finished product should be solid or hollow. Some cosmetic parts may not need to be complete inside, and hollowed out parts may reduce material costs. But the weight effect could have the most valuable impact for some. Both Boeing and Airbus use 3D-printed materials to reduce the weight of their aircraft. The most dramatic outcome of honeycombing is the metal microlattice, a solid metal foam layer that is the lightest metallic structure ever produced. It is 99 percent air, consisting of a 3D open-cell polymer structure made of interconnected hollow tubes. In the future, there may be many applications for this material.

- **Fewer parts for complex geometries**: 3D printers are capable of manufacturing various combinations of metal alloys to satisfy the specifications of the end-product. The opportunity to test different compositions easily and cheaply may give rise to the creation of materials that have not been seen before and may be better suited to comply with certain design specifications than with traditional materials.

One interesting feature of 3DP is that middleware has the ability to bridge the gap between the design and the 3D printer, which means that the middleware will use the design parameters as inputs and, through the algorithms, draw on the optimal mix of source materials for the final product defined in the specification. In addition, it will be possible to produce components with the desired mechanical properties. For example, titanium aluminum (TiAl), used to produce turbine blades, is very brittle at room temperature and is difficult to use with conventional manufacturing techniques, such as casting.

- **Science and technology advancing material management**: Bar-coding significantly improved tracking and positioning of parts along the supply chain. Later, radio frequency identification (RFID) technology allowed parts to be detected via GPS coordinates. 3D-printed parts will change the inventory mix and change the tracking process again. RFID technology will not necessarily be a thing of the past; however, metal powder for printing will be purchased in such large volumes that RFID tags may be the less useful technology for tracing the powder from the source to be used in the production of 3D-printed parts. A research team at Harvard University has extended its micro scale 3DP technology to the fourth dimension: Time. Inspired by nature, 3D-printed particles forms different structures based on the response to environmental stimuli. This is an example of progress in the assembly of programmable materials and enables the ultimate agile supply chain, because the programmable material can be transformed into what is required.

1.4 ACTIVE 3D PRINTING MARKET

The 3DP market for printer unit shipments is projected to grow at a compound annual growth rate (CAGR) of 121.3% by 2019 and to exceed $14.6 billion. Suppliers of 3DP goods and services can be segmented as follows:

- **Major industrial companies**: 3D printing has always been a niche industry, with a few companies leading the field. That said, the space is rising rapidly and is projected to be worth more than $30 billion by 2022. The factors leading to this growth include mass customization, production of complex parts, government investment in 3D printing, and improvements in manufacturing performance. For those interested in jumping into this exciting market, it's worth getting acquainted with the major players.

- **Specialized 3D printer vendors**: The majority of 3D printer manufacturers focus on specialized technologies for the creation of additive layers. Some of the 3D printer companies have been concentrating on industrial applications. The inventory turn for these large units is not very fast, and the unsold inventory can very quickly become obsolete in this space. Some of the major manufacturers are Hewlett-Packard (HP), Proto Labs Autodesk, EOS, Exone, Stratasys, 3D Systems, Reprap, SLM Solutions Group, Organovo, Voxeljet, and Ultimaker.
- **3DP service bureaus**: Some manufacturers of 3D printers also provide after-sales services and consulting services. According to a survey, when businesses plan to outsource a 3D print job to a service office, 34% of respondents find consistency to be the most significant criteria. Some of the choices that businesses have are: Advanced Manufacturing, Aspect, DiSanto Technology (Arcam Unit), Hyphen Services, i.materialise, Ponoko, and Shapeways.
- **Marketplace providers**: Marketplaces are online platforms that serve as intermediaries between individuals or companies who own a 3D printer and users that want to produce 3D objects. They usually provide 3D printing services across a large network of printers in a variety of locations around the world. 3D Hubs, for example, links a file to a printer in one of their 5,000 global hub locations which can be successful for the consumer and increase the accessibility of 3DP to anyone. Other players are Additer, Maker 6, Make XYZ, Materialise, and Spark.
- **Software vendors**: One key differentiator for companies to consider is an open-source platform for crowdsourcing ideas and the provision of completed spare parts blueprints. Companies that have ventured down this avenue have been announced by hobbyists eager to download upgrades or replacement parts. Companies offering open source platforms for products can create brand loyalty and, above all, can ensure that there is a single file in the midst of a legitimate batch of counterfeit reverse-engineered replacement pieces. C, for example, is the first open 3DP software platform (not including the Reprap pioneer, as it was limited to hardware). C was created by Autodesk and runs with a 3D printer from Ember.
- **Professional services providers**: Putting together various consultancy expertise, including planning, supply chain, design and engineering, product creation, technology, analytics, as well as tax and legal services, helps navigate the end-to-end 3D printing cycle and assists companies on their path toward adoption. In an inventory of thousands, the option of which components are ideally suited to 3D printing involves diagnosis and methodology. In addition, they can recommend and bring together the right 3D printing route vendors.

1.5 3D PRINTING PROCESSES

There are many different 3D branded printing processes that can be grouped into seven categories: A total of seven different types of additive manufacturing processes have been established and identified. These seven 3D printing processes brought forth ten different types of 3D printing technology that 3D printers use today.

- 3DPrinting Process: Material Extrusion
  - Fused Deposition Modeling (FDM)
- 3D Printing Process: Vat Polymerization
  - Stereolithography (SLA)
  - Digital Light Processing (DLP)
- 3D Printing Process: Powder Bed Fusion
  - 3D Printing Process: Powder Bed Fusion (Polymers)
    - Selective Laser Sintering (SLS)
  - 3D Printing Process: Powder Bed Fusion (Metals)
    - Direct Metal Laser Sintering (DMLS)/Selective Laser Melting (SLM)
  - Electron Beam Melting (EBM)
- 3D Printing Process: Material Jetting
  - Material Jetting (MJ)
  - Drop on Demand (DOD)
- 3D Printing Process: Binder Jetting
  - Sand Binder Jetting
  - Metal Binder Jetting
- Sheet Lamination
- Directed Energy Deposition

1.5.1 MATERIAL EXTRUSION

Material extrusion is a 3D printing process where a filament of solid thermoplastic material is pushed through a heated nozzle, melting it in the process. The printer deposits the material on a build platform along a predetermined path, where the filament cools and solidifies to form a solid object (Figure 1.1).

- **Types of 3D printing technology**: Fused deposition modeling (FDM), sometimes called fused filament fabrication (FFF)
- **Materials**: Thermoplastic filament (PLA, ABS, PET, TPU)
- **Dimensional accuracy**: ±0.5% (lower limit ±0.5 mm)
- **Common applications**: Electrical housings; Form and fit testings; Jigs and fixtures; Investment casting patterns
- **Strengths**: Best surface finish; Full color and multi-material available
- **Weaknesses**: Brittle, not sustainable for mechanical parts; Higher cost than SLA/DLP for visual purposes

1.5.1.1 Fused Deposition Modeling (FDM)

Material extrusion devices are the most commonly available—and the cheapest—types of 3D printing technology in the world. You might be familiar with them as fused deposition modeling, or FDM. They are also sometimes referred to as fused

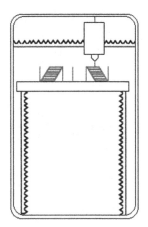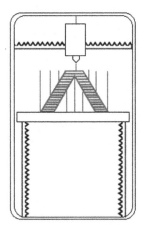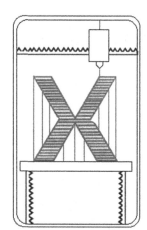

FIGURE 1.1 Material extrusion.

filament fabrication, or FFF. The way it works is that a spool of filament is loaded into the 3D printer and fed through to a printer nozzle in the extrusion head. The printer nozzle is heated to a desired temperature, whereupon a motor pushes the filament through the heated nozzle, causing it to melt. The printer then moves the extrusion head along specified coordinates, laying down the molten material onto the build plate where it cools down and solidifies. Once a layer is complete, the printer proceeds to lay down another layer. This process of printing cross-sections is repeated, building layer-upon-layer, until the object is fully formed. Depending on the geometry of the object, it is sometimes necessary to add support structures, for example if a model has steep overhanging parts.

1.5.2 Vat Photopolymerization

Vat photopolymerization is a 3D printing process where a photo-polymer resin in a vat is selectively cured by a light source. The two most common forms of vat polymerization are SLA (stereolithography) and DLP (digital light processing). The fundamental difference between these types of 3D printing technology is the light source they use to cure the resin. SLA printers use a point laser, in contrast to the voxel approach used by a DLP printer (Figure 1.2).

- Types of 3D printing technology: Stereolithography (SLA);Direct light processing (DLP)
- Materials: Photopolymer resin (standard, castable, transparent, high Temperature)
- Dimensional accuracy: ±0.5% (lower limit ±0.15 mm)
- Common applications: Injection mold-like polymer prototypes; Jewelry (investment casting); Dental applications; Hearing aids
- Strengths: Smooth surface finish; Fine feature details
- Weaknesses: Brittle, not suitable for mechanical parts

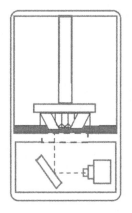 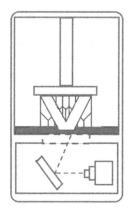 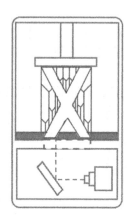

FIGURE 1.2 Vat photopolymerization.

1.5.2.1 Stereolithography (SLA)

SLA holds the historical distinction of being the world's first 3D printing technology. Stereolithography was invented by Chuck Hull in 1986, who filed a patent on the technology and founded the company 3D Systems to commercialize it. An SLA printer uses mirrors, known as galvanometers or galvos, with one positioned on the X-axis and another on the Y-axis. These galvos rapidly aim a laser beam across a vat of resin, selectively curing and solidifying a cross-section of the object inside this build area, building it up layer by layer. Most SLA printers use a solid state laser to cure parts. The disadvantage to these types of 3D printing technology using a point laser is that it can take longer to trace the cross-section of an object when compared to DLP.

1.5.2.2 Digital Light Processing (DLP)

Looking at digital light processing machines, these types of 3D printing technology are almost the same as SLA. The key difference is that DLP uses a digital light projector to flash a single image of each layer all at once (or multiple flashes for larger parts).Because the projector is a digital screen, the image of each layer is composed of square pixels, resulting in a layer formed from small rectangular blocks called voxels. DLP can achieve faster print as compared to SLA. That's because an entire layer is exposed all at once, rather than tracing the cross-sectional area with the point of a laser. Light is projected onto the resin using light-emitting diode (LED) screens or a UV light source (lamp) that is directed to the build surface by a digital micro-mirror device (DMD). A DMD is an array of micro-mirrors that control where light is projected and generate the light-pattern on the build surface.

1.5.3 POWDER BED FUSION

1.5.3.1 Powder Bed Fusion (Polymers)

Powder bed fusion is a 3D printing process where a thermal energy source will selectively induce fusion between powder particles inside a build area to create a solid object. Many powder bed fusion devices also employ a mechanism for applying and smoothing powder simultaneous to an object being fabricated, so that the final item is encased and supported in unused powder (Figure 1.3).

- Types of 3D printing technology: Selective laser sintering (SLS)
- Materials: Thermoplastic powder (Nylon 6, Nylon 11, Nylon 12)
- Dimensional accuracy: ±0.3% (lower limit ±0.3 mm)
- Common applications: Functional parts; Complex ducting (hollow designs); Low run part production
- Strengths: Functional parts, good mechanical properties; Complex geometries
- Weaknesses: Longer lead times; Higher cost than FFF for functional applications

1.5.3.1.1   Selective Laser Sintering (SLS)

Creating an object with powder bed fusion technology and polymer powder is generally known as selective laser sintering (SLS). As industrial patents expire, these types of 3D printing technology are becoming increasingly common and lower cost. First, a bin of polymer powder is heated to a temperature just below the polymer's melting point. Next, a recoating blade or wiper deposits a very thin layer of the powdered material—typically 0.1 mm thick—onto a build platform. A CO_2 laser beam then begins to scan the surface. The laser will selectively sinter the powder and solidify a cross-section of the object. Just like SLA, the laser is focused onto the correct location by a pair of galvos. When the entire cross-section is scanned, the build platform will move down one layer thickness in height. The recoating blade deposits a fresh layer of powder on top of the recently scanned layer, and the laser will sinter the next cross-section of the object onto the previously solidified cross-sections. These steps are repeated until all objects are fully manufactured. Powder which hasn't been sintered remains in place to support the object that has, which eliminates the need for support structures.

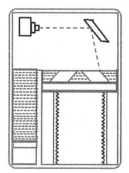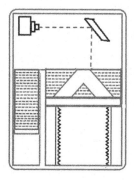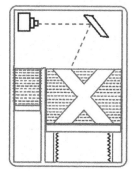

FIGURE 1.3   Powder bed fusion (polymers).

1.5.3.2   Powder Bed Fusion (Metals)

Metal powder bed fusion is a 3D printing process which produces solid objects, using a thermal source to induce fusion between metal powder particles one layer at a time. Most powder bed fusion technologies employ mechanisms for adding powder as the object is being constructed, resulting in the final component being encased in the metal powder. The main variations in metal powder bed fusion technologies come from the use of different energy sources; lasers or electron beams (Figure 1.4).

- **Types of 3D printing technology**: Direct metal laser sintering (DMLS); Selective laser melting (SLM); Electron beam melting (EBM)
- **Materials**: Metal Powder: Aluminum, stainless steel, titanium
- **Dimensional accuracy**: ±0.1 mm
- **Common applications**: Functional metal parts (aerospace and automotive); Medical; Dental
- **Strengths**: Strongest, functional parts; Complex geometries
- **Weaknesses**: Small build sizes; Highest price point of all technologies

1.5.3.2.1   Direct Metal Laser Sintering (DMLS)/Selective Laser Melting (SLM)

Both direct metal laser sintering (DMLS) and selective laser melting (SLM) produce objects in a similar fashion to SLS. The main difference is that these types of 3D printing technology are applied to the production of metal parts.

DMLS does not melt the powder but instead heats it to a point so that it can fuse together on a molecular level. SLM uses the laser to achieve a full melt of the metal powder forming a homogeneous part. This results in a part that has a single melting temperature (something not produced with an alloy).

This is the main difference between DMLS and SLM; the former produces parts from metal alloys, while the latter forms single element materials, such as titanium. Unlike SLS, the DMLS and SLM processes require structural support, in order to limit the possibility of any distortion that may occur (despite the fact that

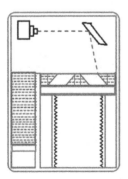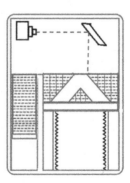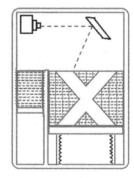

FIGURE 1.4   Powder bed fusion (metals).

the surrounding powder provides physical support). DMLS/SLM parts are at risk of warping due to the residual stresses produced during printing, because of the high temperatures. Parts are also typically heat-treated after printing, while still attached to the build plate, to relieve any stresses in the parts after printing.

1.5.3.2.2 Electron Beam Melting (EBM)

Distinct from other powder bed fusion techniques, electron beam melting (EBM) uses a high energy beam, or electrons, to induce fusion between the particles of metal powder. A focused electron beam scans across a thin layer of powder, causing localized melting and solidification over a specific cross-sectional area. These areas are built up to create a solid object.

Compared to SLM and DMLS types of 3D printing technology, EBM generally has a superior build speed because of its higher energy density. However, things like minimum feature size, powder particle size, layer thickness, and surface finish are typically larger. Also important to note is that EBM parts are fabricated in a vacuum, and the process can only be used with conductive materials.

1.5.4 MATERIAL JETTING

Material Jetting is a 3D printing process where droplets of material are selectively deposited and cured on a build plate. Using photopolymers or wax droplets that cure when exposed to light, objects are built up one layer at a time. The nature of the material jetting process allows for different materials to be printed in the same object. One application for this technique is to fabricate support structures from a different material to the model being produced (Figure 1.5).

- **Types of 3D printing technology**: Material jetting (MJ); Drop on demand (DOD)

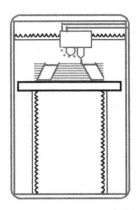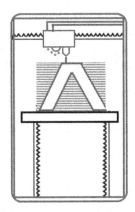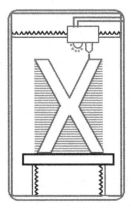

FIGURE 1.5 Material jetting.

- **Materials**: Photopolymer resin (standard, castable, transparent, high temperature)
- **Dimensional accuracy**: ±0.1 mm
- **Common applications**: Full color product prototypes; Injection mold-like prototypes; Low run injection molds; Medical models
- **Strengths**: Best surface finish; Full color and multi-material available
- **Weaknesses**: Brittle, not suitable for mechanical parts; Higher cost than SLA/DLP for visual purposes

1.5.4.1   Material Jetting (MJ)

Material jetting (MJ) works in a similar way to a standard inkjet printer. The key difference is that, instead of printing a single layer of ink, multiple layers are built upon each other to create a solid part. The print head jets hundreds of tiny droplets of photopolymer and then cures/solidifies them using an ultraviolet (UV) light. After one layer has been deposited and cured, the build platform is lowered down one layer thickness and the process is repeated to build up a 3D object. MJ is different from other types of 3D printing technology that deposit, sinter, or cure builds material using point-wise deposition. Instead of using a single point to follow a path which outlines the cross-sectional area of a layer, MJ machines deposit build material in a rapid, line-wise fashion. The advantage of line-wise deposition is that MJ printers are able to fabricate multiple objects in a single line with no impact on build speed. So long as models are correctly arranged, and the space within each build line is optimized, MJ is able to produce parts at a speedier pace than other types of 3D printer. Objects made with MJ require support, which are printed simultaneously during the build from a dissolvable material that's removed during the post-processing stage. MJ is one of the only types of 3D printing technology to offer objects made from multi-material printing and full-color.

1.5.4.2   Drop on Demand (DOD)

Drop on demand (DOD) is a type of 3D printing technology that uses a pair of ink jets. One deposits the build materials, which is typically a wax-like material. The second is used for dissolvable support material. As with typical types of 3D printing technology, DOD printers follow a predetermined path to jet material in a point-wise deposition, creating the cross-sectional area of an object layer-by-layer. DOD printers also use a fly-cutter that skims the build area after each layer is created, ensuring a perfectly flat surface before commencing the next layer. DOD printers are usually used to create patterns suitable for lost-wax casting or investment casting, and other mold-making applications.

1.5.5   BINDER JETTING

Binder jetting is a 3D printing process where a liquid bonding agent selectively binds powder bed regions. Binder jetting is a 3D printing technology similar to SLS, requiring an initial layer of powder on a built-in platform (Figure 1.6).

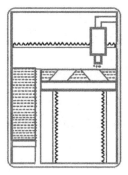 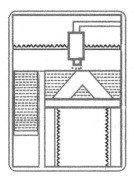 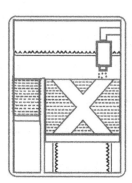

FIGURE 1.6 Binder jetting.

But unlike SLS, which uses a sinter powder laser, binder jetting moves the print head over the powdered surface of the binder droplets, which are typically 80 microns in diameter. These droplets bind the powder particles together to make each layer of the object. Once a layer has been printed, the powder bed is lowered and a new layer of powder spreads over the newly printed layer. This process is repeated until the whole object is formed. The object is then left in the powder to recover and gain energy. Afterward, the object is removed from the powder bed and any unbound powder is removed by compressed air.

- **Types of 3D printing technology**: Binder jetting (BJ)
- **Materials**: Sand or metal powder: Stainless/bronze, full sand, silica (sand casting)
- **Dimensional accuracy**:±0.2 mm (metal) or ±0.3 mm (sand)
- **Typical applications**: Functional metal parts; Full color models; Sand casting
- **Strengths**: Low cost; Large volumes; Functional metal parts
- **Weakness**: Mechanical properties not as strong as metal powder melting beds

1.5.5.1 Sand Binder Jetting

These are low-cost types of 3D printing technology for the production of sandstone or gypsum components with sand binder jetting devices.

For full color models, objects are manufactured using a plaster-based or PMMA powder in conjunction with a liquid binding agent. The print head first jets the binding agent, while the secondary print head jets the color, allowing full color models to be printed. Once the parts have been fully cured, they are removed from the loose unbonded powder and cleaned. Parts are also exposed to infiltrant material to improve mechanical properties. A large number of infiltrants are available, each resulting in different properties. Coatings may also be added to improve the vibrancy of colors. Binder jetting is also useful for the development of sand molds and cores.

Cores and molds are generally sand-printed, although artificial sand (silica) can be used for special applications. After printing, the cores and molds are removed from the construction area and washed to remove any loose sand. Typically, the molds are ready for casting immediately. After casting, the mold is broken apart and the final metal component is removed. The major advantage of manufacturing sand casting cores and molds with binder jetting is that the process can create large, complex geometries at a relatively low cost.

In addition, the process is quite easy to integrate without disruption into existing manufacturing or foundry processes.

1.5.5.2 Metal Binder Jetting

Binder jetting can also be used for the manufacture of metal objects. The metal powder is bound by a polymer binding agent. The production of metal objects using binder jetting enables the production of complex geometries well beyond the capabilities of conventional manufacturing techniques. Functional metal objects can only be produced via a secondary process like infiltration or sintering, however. The cost and quality of the end result generally defines which secondary process is the most appropriate for a certain application. Without these additional steps, a part made with metal binder jetting will have poor mechanical properties.

The infiltration secondary process works as follows: Initially metal powder particles are bound together using a binding agent to form a "green state" object. Once the objects have fully cured, they are removed from the loose powder and placed in a furnace, where the binder is burnt out. This leaves the object at around 60% density with voids throughout. Next, bronze is used to infiltrate the voids via capillary action, resulting in an object with around 90% density and greater strength. However, objects made with metal binder jetting generally have lower mechanical properties than metal parts made with powder bed fusion. The sintering secondary process can be applied where metal parts are made without infiltration. After printing is complete, green state objects are cured in an oven. Next, they're sintered in a furnace to a high density of around 97%. However, non-uniform shrinkage can be an issue during sintering and should be accounted for at the design stage.

1.5.6 SHEET LAMINATION

In the sheet lamination additive manufacturing process, thin sheets of material are bonded together using adhesives or a heat source to form a three-dimensional product. The sheet lamination processes are also known as: Ultrasonic additive manufacturing (UAM) when ultrasonic bonding is used to laminate thermoplastic sheets together and laminated object manufacturing (LOM) when adhesives are used for lamination.

1.5.6.1 Materials Used in Sheet Lamination

Polymers are often used but paper or metal foils are also typically processed and find application in cases where heat sensitive materials cannot be used and low costs must be realized. Almost any polymer can be used as long as it is available in thin sheet form and can be bonded by either adhesives or heat.

The main advantages of sheet lamination are:

i. Low materials cost
ii. Many substrates are available (e.g. paper, film, foil)
iii. Process does not require a closed environment
iv. High volumetric build rates
v. Allows for combination of materials and embedding components

The key disadvantages are that complex geometries are difficult to manufacture and may be less precise than other additive manufacturing methods. Many characteristics of this method are as follows:

i. Using binding materials such as adhesives or energy (e.g. ultrasonic welding)
ii. Relatively large components may be produced
iii. Possibility of low cost, easily accessible building materials such as paper, plastic film, or metal foil
iv. Bonding equipment can be easy (even by hand) or automated
v. Significant applications: Large parts, tooling

1.5.7 DIRECTED ENERGY DEPOSITION

Directed energy deposition methods usually do not use polymeric materials but use metal wire or powder. High energy sources, such as the laser, are directed at the material to melt and build up the product. Directed energy deposition is assumed to be a more complex and expensive additive manufacturing process, but is generally used to repair or add extra materials to existing components. Other characteristics of directed energy deposition include: Similar to powder bed fusion except material is first injected into an energy field. Common substrates are metal, metal wire, glass, and ceramics.
Strengths:

i. Can operate in open air
ii. Multiple materials can be used
iii. Large parts are possible
iv. High single point deposition rates
v. Not limited by direction or axis

Limitations:

i. Expensive equipment, lower resolutions, and reduced ability to manufacture complex parts
ii. Final machining is often required

Major applications: Repair or build-up of high volume parts (Table 1.1).

TABLE 1.1

Recap and Description of These Processes

Process	Description	Technology
Photo polymerization	A vat of liquid photopolymer resin is cured by selective exposure to light (through a laser or projector). This then initiates polymerization and transforms the exposed areas into a solid component.	Stereolithography (SLA) Digital light processing (DLP) Continuous liquid interphaseproduction (CLIP) Scan, spin, and selectively photocure (3SP)
Material jetting	Material droplets are deposited layer by layer to produce objects. Common varieties include jetting and curing of photocurable resins with UV light, as well as jetting of thermally molten materials, which then solidify at ambient temperature. This process was the origin of the term "3D printing."	3D printing (3DP) Multi-jet modeling (MJM) Drop on demand (DOD)
Binder jetting	Bonding agents are applied selectively to thin layers of powdered material to form parts layer by layer. Binders shall contain organic and inorganic materials. Metal or ceramic powdered parts are commonly fired in the furnace after they have been printed.	Drop on powder (DOP) Powder bed printing
Material extrusion	The material is extruded by a nozzle or an orifice in tracks or beads, which are then integrated into multi-layer models. Popular varieties include hot thermoplastic extrusion (similar to a hot glue gun) and syringe dispensing.	Fused deposition modeling (FDM) Fused filament fabrication (FFF)
Powder bed fusion	Powdered materials are selectively deposited by melting together using a heat source such as a laser or electron beam. The powder surrounding the consolidated component works as a support material for the overhanging materials.	Selective heat sintering (SHS) Direct metal laser sintering (DMLS) Electron beam melting (EBM) Selective laser melting (SLM) Selective laser sintering (SLS)
Sheet lamination	Material sheets are stacked and laminated with each other to form an object. The method of lamination can be adhesive, ultrasonic welding, or brazing (metals). Excessive regions are cut layer by layer and removed after the object is built.	Selective deposition lamination (SDL) Laminated object manufacturing (LOM) Ultrasonic additive manufacturing (UAM)
Direct energy deposition	Metal powder or wire is feed into a melt pool formed on the surface of the part where it is attached to the underlying part or layer. The source of energy is normally a laser beam or an electron beam. In essence, this process is a type of automated build-up welding.	Laser metal deposition (LMD) Electron beam free-form fabrication (EBF3) Direct metal deposition (DMD) Laser engineered net shaping (LENS)

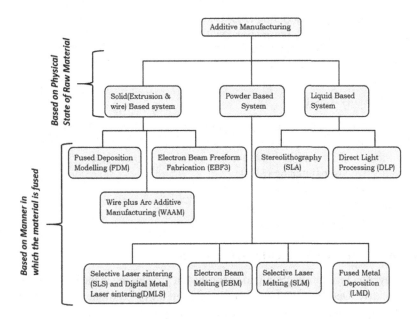

FIGURE 1.7 Classification of additive manufacturing systems.

1.6 CLASSIFICATION OF ADDITIVE MANUFACTURING SYSTEMS

A better way to broadly classify AM systems is by the initial form of their material; all AM systems can be easily categorized into:

A) Liquid-based
B) Solid-based
C) Powder-based

1.6.1 LIQUID-BASED

Building materials are in liquid state and the following additive manufacturing systems fall under this category (Figure 1.7):

  i. Stereolithography apparatus (SLA)
 ii. PolyJet 3D printing
iii. Multijet printing (MJP)
 iv. Solid object ultraviolet-laser printer (SOUP)
  v. Rapid freeze prototyping

1.6.2 SOLID-BASED

The building material is in solid state (excluding powder). The solid type may include the shape of the wire, rolls, laminates, and pellets. The following AM systems fall under this category:

  i. Fused deposition modeling (FDM)
 ii. Selective deposition lamination (SDL)
iii. Laminated object manufacturing (LOM)
 iv. Ultrasonic consolidation

1.6.3 POWDER-BASED

Building material is powder (a type of grain). All powder-based AM systems use the joining/binding method. The following AM systems fall under this category:

  i. Selective laser sintering (SLS)
 ii. ColorJet printing (CJP)
iii. Laser engineered net shaping (LENS)
 iv. Electron beam melting (EBM), etc.

1.7 ADVANTAGES AND LIMITATIONS

1. Layer by layer production makes the design process much more flexible and creative. Designers no longer need to design for production; rather they can build a component that is lighter and stronger by means of better design. Parts can be fully re-designed in such a way that they are stronger in the places they need to be and lighter overall.
2. 3D printing significantly accelerates the design and prototyping process. There's no problem creating one part at a time, and changing the design every time it's produced. Parts can be created in hours. The design time is taken down to days or weeks, opposed to months. Also, as the price of 3D printers has decreased over the years, some 3D printers are now within the financial reach of ordinary consumers or small companies.
3. In general, the drawbacks of 3D printing include costly hardware and expensive materials. This leads to expensive parts, making it difficult to compete with mass production. A CAD designer is also needed to create what the consumer has in mind and can be costly if the component is very complex.
4. 3D printing is not the solution for everyone involved in the production method; however, its improvement is helping to accelerate design and engineering more than ever before. Through the use of 3D printers, designers are able to create one piece of art, intricate building and product designs, and make parts while in space!
5. Many industries are beginning to see the impact of 3D printing. There have been articles saying that 3D printing will bring about the next industrial revolution, by returning a means of production back within reach of the designer or the consumer.

1.8 ADDITIVE VS. CONVENTIONAL MANUFACTURING PROCESSES

The growing success of additive manufacturing is due to its advantages over conventional manufacturing. However, these strengths often come along with certain weaknesses. The weaknesses provide opportunities for corrective action through the development of new polymeric materials.

Strengths:

- Elimination of design limitations
- Enable parts to be generated with complicated geometries: Honeycomb structures, cooling channels, etc., and no additional costs associated with complexity
- Build speed; Reduce lead time
- Flexibility of design
- No expensive tooling criteria
- Dimensional accuracy
- Broad variety of materials (polymers, metals, and ceramics)
- Well suited for the manufacture of high-value replacement and repair parts
- Green production, clean, minimal waste
- Small footprint for manufacturing and constantly rising equipment costs

Limitations:

- Surface roughness
- Low density, porosity
- Lack of data on the end-use properties of parts to be produced (e.g. thermal and chemical stability, strength, etc.)
- Limited to relatively small parts

1.9 APPLICATIONS

The development of innovative, advanced additive manufacturing technologies has progressed quickly yielding broader and high value applications. This accelerating trend has been due to the benefits of additive manufacturing compared to more conventional manufacturing processes. Some of these benefits are:

1. Lower energy consumption
2. Less waste
3. Less dedicated tooling
4. Reduced development costs and time to market
5. Innovative designs and geometries
6. Part consolidation (fewer parts with more complex design)
7. Customization of parts (e.g. for medical implants, specialty repair parts, parts where other manufacturing facilities are not available such as on ships or in space—Table 1.2).

1.10 EXERCISES

1. Discuss the development of additive manufacturing and 3D printing.
2. What are various major trends shaping the evaluation of 3D printing?

TABLE 1.2

Industries to Benefit Most in the Immediate Future from Additive Manufacturing and the Value Provided

Industry	Applications and Value
	Aerospace and Defense i. Concept modeling and prototyping ii. Manufacturing of low-volume specific components (electronics, engine parts, etc.) iii. Manufacturing of spare parts everywhere iv. Manufacture of structures using lightweight, high strength materials
	Automotive i. Testing the design of the part to check correctness and completeness ii. Parts for racing vehicles, luxury sports cars, vintage cars, etc. iii. Replacement of parts that are defective or that cannot be purchased iv. Manufacture of structures using lightweight, high strength materials
	Electronics i. Embedding radio frequency identification (RFID) systems embedded in solid materials ii. Short lead time for electronic goods iii. Three-dimensional micro-electromechanical systems based on polymer iv. Microwave circuits built on paper substrates
	Tool and Mold Making i. Universal device holder with standard pocket sizes ii. Die casting forms iii. Injection molding devices iv. Tools for prototyping short-term surgical systems
	Medical i. Design and modeling of customized implants and medical devices ii. Processes for the manufacture of "smart scaffolds" and the construction of 3D biological and tissue models

3. Discuss the technological improvement in additive manufacturing.
4. What are the various 3DP processes?
5. Classify the additive manufacturing systems.
6. Discuss the advantages of additive manufacturing.
7. 7.What are the differences between additive and conventional manufacturing processes?
8. What are the various applications of 3DP?

1.11 MULTIPLE-CHOICE QUESTIONS

1. Additive manufacturing uses much less material than other subtractive manufacturing processes.
   a) True
   b) False
   Ans: (a)

2. Which of the following is typically the cheapest type of 3D printer?
   a) FDM
   b) SLA
   c) Powder-based
   d) SLM
   Ans: (a)

3. SLA printing uses a plastic strand that's pushed through a heated nozzle.
   a) True
   b) False
   Ans: (b)

4. 3D printing technology is expanding and is now able to print metal parts.
   a) True
   b) False
   Ans: (a)

5. Which of the following is typically the most expensive type of 3D printer?
   a) SLA
   b) SLM
   c) FDM
   d) None of the above
   Ans: (b)

6. FDM stands for fused deposition modeling.
   a) True
   b) False
   Ans: (a)

7. SLA stands for stereolighting amplification.
   a) True
   b) False
   Ans: (b)

8. FDM printers can print multiple materials at one time.
   a) True
   b) False
   Ans: (a)

9. Which type of printer uses an enclosed build area?
    a)   SLA
    b)   SLS
    c)   MDS
    d)   FDM
    Ans: (a)

10. What printer melts metal?
    a)   SLS
    b)   SLM
    c)   SLA
    d)   FDM
    Ans: (b)

11. SLA printer's package material is in a...
    a)   Chain
    b)   Spool
    c)   Cartridge
    d)   None of the above
    Ans: (c)

12. Which type of 3D printer uses a pool of resin to create the solid part?
    a)   FDM
    b)   SLA
    c)   SNL
    d)   None of the above
    Ans: (b)

13. What material is not used in 3D printing?
    a)   Nylon
    b)   ABS
    c)   PLA
    d)   PVC
    Ans: (d)

14. When printing a part on an FDM printer, the model should be oriented such that the layers are parallel to the direction of stresses that will be placed on the part.
    a)   True
    b)   False
    Ans: (b)

15. Which should be considered when orienting the part on the build plate in the slicing software?
    a)   Holes should always be printed horizontally

b) The footprint of the part should be as small as possible
c) You should minimize the number of overhangs
d) All of the above
Ans: (c)

16. Which of the following does NOT influence how refined the 3D printed part will be?
    a) Layer thickness
    b) Using support material
    c) Part orientation
    d) All the above
    Ans: (d)

17. Which file type is most commonly exported from CAD software?
    a) SLDRT
    b) JPG
    c) STL
    d) X3G
    Ans: (c)

18. FDM build plates are prepared by...
    a) Putting hair spray on it
    b) Putting a layer of painters tape on it
    c) Putting a glue stick layer on it
    d) All the above
    Ans: (d)

19. What does SLS stand for?
    a) Selective laser sintering
    b) Selective lithographic solution
    c) Separated light sintering
    d) None of the above
    Ans: (a)

2 CAD for Additive Manufacturing

2.1 INTRODUCTION

It is clear that additive manufacturing would not have been possible without computers and have been developed so far if it were not for the development of 3D solid modeling CAD (computer-aided design). The quality, reliability, feasibility, and ease of use of 3D CAD meant that virtually any geometry could be modeled which enhanced our design capability. Some of the most impressive models made using additive manufacturing are those that illustrate the ability to manufacture complex shapes in a single stage without the need to assemble or use secondary tools. Virtually every commercial solid modeling CAD system is capable of output to an AM machine. This is because, in most cases, the only information required by the AM machine from the CAD system is the external geometric form. The machine is not required to know how the part was modeled, any of the features, or any of the functional elements. The part can be built as long as the external geometry can be defined.

2.2 PREPARATION OF CAD MODELS: THE STL FILE

The first step in each product design process is to visualize and conceptualize the function and appearance of the product. This can take the form of text information, sketches, and three-dimensional computer models. As far as the process chain is concerned, the first enabler of AM technologies is the 3D digital computer-aided design (CAD) models where the conceptualized product exists in the "computer" space and the values of its geometry, material, and properties are stored in digital form and are easily retrievable.

Generally, AM process chains start with 3D CAD modeling. The process of creating a 3D CAD model from a concept in the designer's mind can take several forms, but all of them involve CAD software programs. The details of these programs and the technology behind them do not fall within the scope of this text, but these programs are a critical enabler of the designer's ability to generate a 3D CAD model that can serve as the starting point for the AM process chain. There is a large number of CAD programs with different modeling concepts, capacities, accessibility, and cost.

Examples include Autodesk Inventor, SolidWorks, CREO, NX, etc. Once a 3D CAD model is produced, steps can be taken in the AM process chain. Although the process chain generally progresses in one direction starting with CAD modeling and ending with a finished part or prototype, it is often an iterative process where changes to the CAD model and design are made to reflect feedback from each step of the process chain. Special to metal powder bed technology, critical feedback may come from geometry and anisotropic properties on parts due to build orientation, distortion of parts, or features due to thermal background of construction, problems with the generation, and removal of support structures, etc. Issues such as these may occur in the AM process chain and may entail changes in design and revisions.

The STL file is derived from the term stereolithography, which was the first commercial AM process developed by the US company 3D Systems in the late 1980s, although some suggested that STL could stand for stereolithography tessellation language. The STL files are created from 3D CAD data on the CAD system. Output is a boundary representation that is approximated by a triangle mesh.

2.2.1   STL FILE FORMAT, BINARY/ASCII

Almost all of the AM technology available today uses the stereolithography (STL) file format. The STL format of the 3D CAD model captures all surfaces of the 3D model by stitching triangles of different sizes on its surfaces. The spatial locations of the vertices of each triangle and the vectors that are normal to each triangle, when combined, allow AM's pre-process programs to determine the spatial locations of the surfaces of the part in the build envelope, and on which side of the surface is the interior of the part.

Although the STL format has been considered a de facto standard, there are inherent limitations to the fact that only geometry information is stored in these files, while all other information that the CAD model may contain is deleted. Information such as the unit, color, material, etc., can play an important role in the functionality of the built-in component being lost during a file translation process. As such, the functionality of the finished parts is limited. The "AMF" format has been developed specifically to address these issues and limitations and is now the standard ASTM/ISO format.

Beyond geometry information, this file format also includes dimensions, color, material, and additional information. Currently, the prevailing file format used by AM systems and supported by CAD modeling programs is still the STL format. Increasing numbers of CAD program companies, including plenty of major programs, have included support for AMF file formats. Currently, the actual use of the information stored in the AMF file is still limited due to the capabilities of the current AM systems and the state of the art technology. The STL file consists of a number of triangular facets.

Each triangular facet is uniquely defined by a normal vector unit and three vertices or corners. The normal vector unit is a line that is perpendicular to the triangle and has a length equal to 1.0. The length of the unit could be in millimeters or inches and is stored using three numbers corresponding to its vector coordinates. The STL file itself does not hold any dimensions, so the AM machine operator must know whether the dimensions are millimeters, inches, or any other unit. Since each vertex also has three numbers, there are a total of 12 numbers for each triangle. The following file shows a simple ASCII STL file describing the right-angled, triangular pyramid structure as shown in Figure 2.1.

```
solid triangular_pyramid
        facet normal  0.0 -1.0 0.0
            outer loop
                vertex 0.0 0.0 0.0
                vertex 1.0 0.0 0.0
                vertex 0.0 0.0 1.0
            endloop
        endfacet
        facet normal  0.0 0.0 -1.0
            outer loop
                vertex 0.0 0.0 0.0
                vertex 0.0 1.0 0.0
                vertex 1.0 0.0 0.0
            endloop
        endfacet
        facet normal  0.0 0.0 -1.0
            outer loop
                vertex 0.0 0.0 0.0
                vertex 0.0 0.0 1.0
                vertex 0.0 1.0 0.0
            endloop
        endfacet
        facet normal  0.577 0.577 0.577
            outer loop
                vertex 1.0 0.0 0.0
                vertex 0.0 1.0 0.0
                vertex 0.0 0.0 1.0
            endloop
        endfacet
    endsolid
```

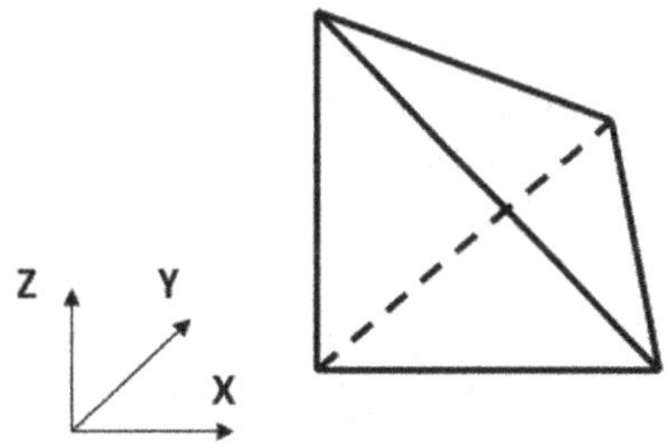

FIGURE 2.1   A right-angled triangular pyramid as defined in the STL sample file.

Note that the file starts with an object name defined as a solid. Triangles can be identified in any order, each as a facet. The facet line also includes the normal vector for the triangle. Note that this normal is calculated from any convenient location on the triangle and may be from one of the vertices or from the center of the triangle. It is defined that the normal is perpendicular to the triangle and the length of the unit. In most systems, the normal is used to define the outer surface of the solid, essentially pointing out the outer surface. The group of three vertices defining the triangle is defined by the terms "outer loop" and "end loop."

The outside of the triangle is better established by the right-hand rule approach. As we look at the triangle from the outside, the vertices should be in the clockwise order. Using the right hand with the thumb pointed upward, the other fingers curl in the direction of the vertical order, the starting vertex being arbitrary. This method is becoming more popular because it avoids having to make any calculations for an additional number (i.e. facet normal) and thus STL files might not even allow the normal to avoid ambiguity.

Note that the bottom left-hand corner is the same as the origin and that each vertex coming out of the origin is of unit length.

A binary STL file can be described as follows:

- The 80-byte ASCII header that might be used to describe the component
- A long integer of 4 bytes that indicates the number of facets in the object
- A list of facets, each 50 bytes long

The facet record will be described in the following manner:

- Three floating values of 4 bytes each for the normal vector
- Three floating values of 4 bytes each for the first vertex
- Three floating values of 4 bytes each for the second vertex
- Three floating values of 4 bytes each for the third vertex
- One unsigned integer of 2 bytes, which should be zero, used for checking

2.2.1.1 Format Specifications

The STL file is a list of facet info. Each facet is uniquely identified by a normal unit (a line perpendicular to the triangle and a length of 1.0) and three vertices (corners). Normal and each vertex are described by three coordinates each, so a total of 12 numbers are stored for each facet.

Facet orientation: The facet defines the surface of a three-dimensional object. As such, each facet is part of the boundary between the inside and the outside of the object. The orientation of the facets (which path is "out" and which path is "in") is given redundantly in two ways that must be consistent. The direction of the normal is outward, first. Second, the vertices are seen in the anti-clockwise order when looking at the object from the outside (right-hand rule). These rules are shown in Figure 2.2.

Vertex-to-vertex rule: Each triangle must share two vertices with each of its neighboring triangles. In other words, the vertex of one triangle cannot lie on the

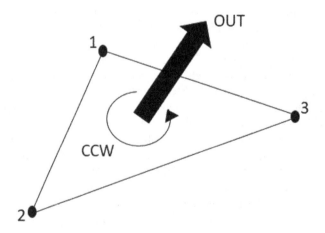

FIGURE 2.2   The orientation of the facet shall be determined by the direction of the normal unit and the order in which the vertices are indicated.

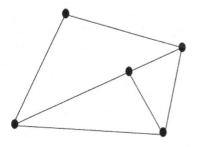 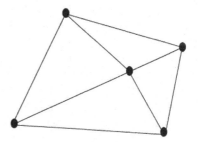

FIGURE 2.3 Vertex-to-vertex rule. Figure (a) shows a violation of Rule (a). The correct configuration is shown in Figure (b).

other side. This is shown in Figure 2.3. The object shown must be placed in the all-positive octant. In other words, all vertex coordinates must be positive-defined (non-negative and non-zero) numbers. The STL file does not contain any scale information; the coordinates are stored in arbitrary units. The official 3D Systems STL specification document states that there is provision for the inclusion of "special attributes for building parameters" but that there is no format for the inclusion of these attributes. Also, the document specifies the details for the "minimum length of the triangle side" and "maximum triangle size," but these numbers are of doubtful significance. Triangle sorting in ascending z-value order is recommended, but not required, in order to optimize the performance of the slice program. Typically, the STL file is saved with the "STL" extension, case-insensitive. This extension may be provided by the slice system or it may allow a specific extension to be defined.

The STL standard includes two ASCII and binary data formats. These are described below separately.

2.2.1.2 STL ASCII Format

The ASCII format is primarily utilized for the development of new CAD interfaces. The large size of its files makes it infeasible for general use.

The syntax of the ASCII STL file is as follows:

solid *name*
$$\left\{ \begin{array}{l} \text{\textbf{facet normal}} \ n_i \ n_j \ n_k \\ \quad \text{\textbf{outer loop}} \\ \qquad \text{\textbf{vertex}} \ v1_x \ v1_y \ v1_z \\ \qquad \text{\textbf{vertex}} \ v2_x \ v2_y \ v2_z \\ \qquad \text{\textbf{vertex}} \ v3_x \ v3_y \ v3_z \\ \quad \text{\textbf{endloop}} \\ \text{\textbf{endfacet}} \end{array} \right\} +$$
endsolid *name*

Boldface indicates the keyword; it must appear in lowercase letters. Note that there is space in "facet normal" and "outer loop," while there is no space in any of the keywords starting with "end." The indentation must be with spaces; the tabs are not allowed. The notation"{...} +" implies that the contents of the brace brackets can be repeated one or more times. Italic symbols are variables to be replaced by user-specific values. The numerical data for facet normal and vertex lines are single precision floats, e.g. 1,23456E+789. A facet normal coordinate may have a leading minus sign; it may not have a vertex coordinate.

2.2.1.3 STL Binary Format

The binary format uses IEEE integer and floating point numeric representation. The syntax for the STL binary file is as follows:

Bytes	Data type	Description
80	ASCII	Header. No data significant
4	Unsigned long integer	Number of facts in file
4	float	*i* for normal
4	float	*j*
4	float	*k*
4	float	*i* for vertext 1
4	float	*y*
4	float	*z*
4	float	*i* for vertext 2
4	float	*y*
4	float	*z*
4	float	*i* for vertext 3
4	float	*y*
4	float	*z*
2	Unsigned integer	Attribute byte count

The notation"{...} +" implies that the contents of the brace brackets can be reproduced one or more times. The syntax of the attribute is not recorded in the formal specification. It is defined that the byte count attribute should be set to zero.

2.2.2 CREATING STL FILES FROM A CAD SYSTEM

Nearly all geometric solid modeling CAD systems can generate STL files from a valid, fully enclosed solid model. Most CAD systems can quickly tell the user if a model is not a solid. This test is particularly necessary for systems that use surface modeling techniques, where it can be possible to create an object that is not fully closed off. Such systems would be used for graphics applications where there is a need for powerful manipulation of surface detail rather than for engineering detailing. Solid modeling systems, like SolidWorks, may use surface modeling as part of the construction process, but the final result is always a solid that would not require such a test.

Most CAD systems use the "Save as" or "Export" feature to convert a native format to an STL file. Typically, there has been some control over the size of the triangles to be used in the model. Since STL uses flat surfaces to approximate curved surfaces, the larger the triangles, the looser the approximation becomes. Most CAD systems do not directly limit the size of the triangles, since it is also clear that the smaller the triangle, the larger the resulting file for the given object.

An effective approach would be to minimize the offset between the triangle and the surface to be represented. The ideal cube with perfectly sharp edges and points can be described by 12 triangles, all with an offset of 0 between the STL file and the original CAD layout. However, few designs would be as convenient and it is necessary to maintain a good balance between surface approximation and an overly large file. Figure 2.4 demonstrates the effect of adjusting the offset triangle parameter in

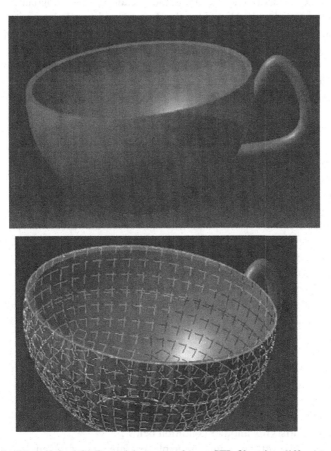

FIGURE 2.4 The original CAD model converted to an STL file using different offset height (cusp) values, showing how the accuracy of the model changes according to the offset triangle.

the STL code. The appropriate value of the required offset would be largely dependent on the resolution or accuracy of the AM process to be used. If the offset is smaller than the basic process resolution, making it smaller will have no effect on the accuracy of the resulting model. Since many AM processes operate around a 0.1 mm layer resolution, a triangle offset of 0.05 mm or slightly lower will be acceptable for manufacturing most of the parts.

2.2.3   CALCULATION OF EACH SLICE PROFILE

Practically every AM system will have the ability to read both binary and ASCII STL files. As most AM processes operate by adding layers of the material of the specified thickness, beginning at the bottom of the part and moving upward, the definition of the part of the file must be processed in order to obtain the profile of each layer.Each layer can be assumed a plane in a nominal Cartesian XY frame. The incremental movement of each layer can then be along the orthogonal axis of the Z. The XY plane, which is located along the Z-axis, can be called a cutting plane. Any triangle intersecting this plane may be considered to contribute to the profile of the slice. An algorithm such as the one in Flowchart 2.1 can be used to extract all the profile segments of the STL file. The outcome of this algorithm is a set of intersecting lines that are ordered according to a set of intersecting planes.

A program that is written as per this algorithm would have a number of additional components, including a way to define the start and end of each file and each plane. Furthermore, there would be no order for each line segment to be defined in terms of the XY components and indexed to the plane corresponding to each Z value. Also, the assumption is that the STL file has an arbitrary set of triangles which are randomly distributed. It may be possible to pre-process each file so that searches can be performed in a more efficient manner. One way to optimize the search would be to order the triangles according to the minimum value of Z. A simple check for the intersection of a triangle with a plane will be to check the value of Z for each vertex. If the Z value of any vertex in the triangle is less than or equal to the Z value of the plane, that triangle may intersect the plane. Using the above test, once it has been established that the triangle may not intersect with the cutting plane, then every other triangle is known to be above that triangle and therefore does not require a check.A similar test could be performed with the maximum value Z of the triangle.

There are a number of discrete scenarios illustrating the intersection of each triangle with the cutting plane:

1. All the vertices of the triangle are above or below the intersecting plane. This triangle is not going to contribute to the profile on this plane.
2. A single vertex is directly on the plane. In this case, there is one intersection point that can be neglected, but the same vertex will be used in other triangles satisfying another condition below.
3. There are two vertices lying on the plane. Here, one of the edges of the corresponding triangle is on that plane, and that edge contributes fully to the profile.

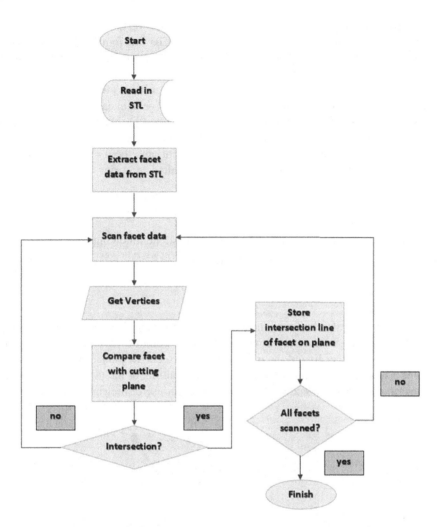

FLOWCHART 2.1 Algorithm used to test triangles and create line intersections. The result is an unordered matrix of intersecting lines.

4. There are three vertices lying on the plane. In this case, the entire triangle contributes entirely to the profile, unless there are also one or more adjacent triangles on the plane, in which case the included edges can be neglected.
5. Another vertex is above or below the intersecting plane, and the other two vertices are on the opposite side of the plane. In this case, the intersection vector must be calculated from the edges of the triangle.

Most triangles should adhere to Scenario 1 or Scenario 5. Scenarios 2–4 may be considered as special cases and require special treatment. Assuming that we have carried out the appropriate checks and that the triangle corresponds to Scenario 5, then we have to take action and generate the corresponding intersecting profile vector.

In this case, there will be two vectors defined by the vertices of the triangle, and these vectors will intersect with the cutting plane. The line joining these two intersection points will form part of the plane's outline. The problem to be resolved is a classical line intersection with a plane problem.

In this case, the line is defined using Cartesian coordinates in (x, y, z). The plane is defined in (x, y) for a specific constant height, z. In a general case, we can therefore project the line and plane onto the x = 0 and y = 0 planes. For the y = 0 plane, we can obtain something similar to Figure 2.5. Points P1 and P2 correspond to two points of the intersecting triangle. Pp is the projected point onto the y = 0 plane to form a unique right-angled triangle. The angle θ can be calculated from

$$\tan\theta = \frac{(Z2 - Z1)}{(X2 - X1)} \tag{2.1}$$

Since we know the z height of the plane, we can use the following equation:

$$\tan\theta = \frac{(Zi - Z1)}{(Xi - X1)} \tag{2.2}$$

and solve for xi.

A point yi can also be found after projecting the same line onto the x = 0 plane to fully define the intersecting point Pi. A second point of intersection can be calculated using another line of the triangle that intersects the plane. These two points will be a line on the plane that forms part of the outline of the model. It is possible to determine the directionality of this line by exact use of the right-hand rule, thus

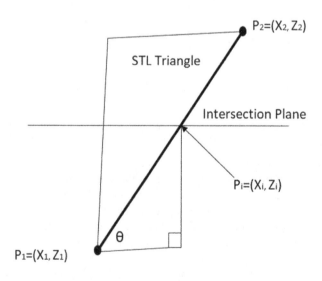

FIGURE 2.5 A vertex taken from an STL triangle projected onto the Y = 0 plane. Since the height Z_i is known, we can derive the intersection point X_i. A similar case can be done for Y_i in the X = 0 plane.

turning this segment of the line into a vector. This can be useful in deciding whether the completed curve is part of the enclosing outline or corresponds to a hole. After all, the intersection lines have been calculated according to Flowchart 2.1; these lines must be joined together to make complete curves. This would be done using an algorithm based on the one described in Flowchart 2.2. In this case, each segment of the line is checked to determine which segment is nearest to it.

The "nearest point" algorithm is necessary because the calculations may not precisely locate points together, even though the same line would normally be used to determine the starting position of one segment and the end of another. Note that this algorithm should actually have a further nesting to test whether a curve has been completed. If the curve is complete, any other remaining line segments would correspond to additional curves. These additional curves could form a nest of curves lying inside or outside others, or they could be separated. The two algorithms listed here concentrate on the intersection of the triangular facet with the cutting surface.

The normal vectors of each triangle could be used to further develop these algorithms. In this way, the actual path of the curve could be determined. It will help to

FLOWCHART 2.2 Algorithm for ordering the intersections of the line into complete outlines. This assumes that there is only one contour in each plane.

assess the nested curves. Outside the section, the outermost curve will point. When the curve set is pointing inward to itself, then it is obvious that there must be a further curve enveloping it (see Figure 2.5). The use of normal vectors may also be helpful in organizing curve sets that are very close to each other.

Once this stage has been completed, there will be a file containing an ordered set of vectors to trace the complete outlines corresponding to the intersecting plane. The way these outlines are used depends somewhat on which AM technology is to be used. Many machines can use the vectors generated in Flowchart 2.2 to control the plotting process for drawing the outlines of each layer. However, most computers will still have to fill in these outlines to make them solid. Flowchart 2.3 uses

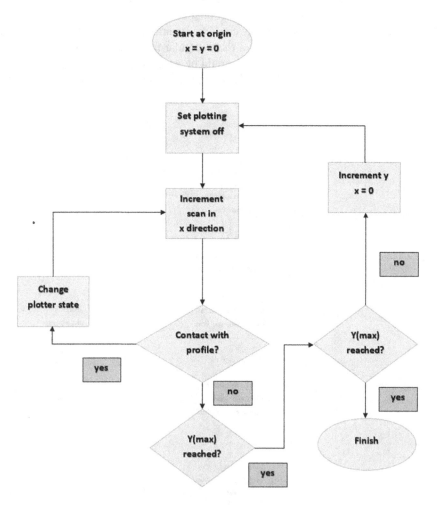

FLOWCHART 2.3 Algorithm for filling a 2D profile based on vectors created using Flowchart 2.2 and a raster scanning method. Assume that the profile suits within the build volume, the raster scans in the X direction, and the lines increase in the Y direction.

an inside/outside algorithm to determine when to switch on a filling mechanism to draw scanning lines perpendicular to one of the plane axes. The assumption is that the part is fully enclosed within the build envelope and therefore the default fill is switched off.

2.2.4 TECHNOLOGY-SPECIFIC ELEMENTS

Flowcharts 2.1, 2.2, and 2.3 are basic algorithms of a standardized type. These algorithms need to be optimized to avoid errors and customize them to fit a specific method. Many refinements can be used to speed up the cutting cycle by reducing duplication, for example. Most AM systems need components designed using support structures. Supports are usually a loose-woven lattice pattern of material placed below the area to be supported. Such a lattice pattern may be a plain square pattern or something more complex like a hexagonal or even a fractal grid. In addition, the lattice could be connected to a part with a tapered area that could be more convenient to remove when compared to thicker connecting edges.

The identification of the regions to be assisted can be made by evaluating the angle of the standard triangle. Those normals that point downward at some previously specified minimum angle will need help. Those triangles that are sloping above that angle will not need support. The supports are extended until they converge either with the base platform or with another upward-facing surface of the portion. Supports attaching to the surface facing upward can also have a taper that allows quick removal. The method that would usually be used would be to stretch the supports from the entire build platform and eliminate any supports that do not overlap with the component at a minimum angle or less (see Figure 2.6).

Support structures could be generated directly as STL models and can be implemented into the already mentioned slicing algorithms. However, they are more likely to be directly created by the proprietary algorithm in the slicing process. Other processing specifications that would rely on different AM technologies include:

Raster scanning: Although many systems will use basic raster scans for each sheet, there are alternatives. Some systems use a switchable raster scan, scan in the X direction of the XY plane for a single layer, and then move to the Y direction for alternate layers. Many systems subdivide the filling area into smaller square regions and use switchable raster scans between squares.

Patterned vector scanning: The material extrusion technology requires a filling pattern to be produced within the enclosed boundary. This is done by using vectors generated using a patterning strategy. In the case of a particular layer, the pattern would be determined by choosing a specific angle for the vectors to travel. The fill is then a zigzag pattern in the direction defined by this angle. Once a zigzag has reached its end, further zigzag fills may be needed to complete a layer (see Figure 2.7 for an example of a zigzag scan pattern).

Hatching patterns: Sheet lamination processes, such as the Helisys LOM and the Solid Center system from Kira, involve material surrounding the part to be hatched with a pattern that enables it to be de-coated once the part has been completed (see Figure 2.8).

FIGURE 2.6   Supports generated for a part build.

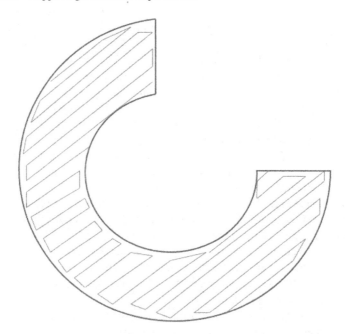

FIGURE 2.7   A scan pattern by means of vector scanning in material extrusion. Note the outline drawn first followed by a small number of zigzag patterns to fill the space.

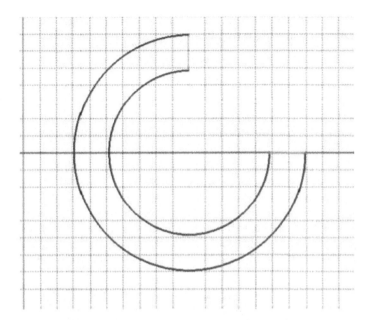

FIGURE 2.8   Hatching pattern for LOM-based (sheet lamination) processes. Note the outer hatch pattern that results in cubes that will be separated from the solid component during post-processing.

2.3   PROBLEMS WITH STL FILES

Although the STL format is quite simple, there may still be errors in the files resulting from the conversion to CAD. Typical problems that may occur in bad STL files are as follows:

Unit changing: This is not strictly the result of a bad STL file. Since US machines still use imperial measurements and most of the rest of the world uses metric, some files may appear scaled because there is no explicit reference to the units used in the STL format. If the person constructing the model is unaware of the purpose of the part, they may construct it approximately 25 times too large or too small in one direction. In addition, the units must correspond to the position of the origin of the system to be used. This typically means that the physical origin of the system is in the bottom left-hand corner, so that all triangle coordinates inside the STL file must be positive. However, this may not be the case for a particular part of the CAD system and therefore some adjustment of the STL file may be required.

Vertex-to-vertex rule: Each triangle must share two of its vertices with each triangle adjacent to it. This means that the vertex cannot intersect the other side, as shown in Figure 2.9. This is not something explicitly stated in the description of the STL file and therefore the generation of the STL file may not comply with this rule. However, a variety of tests can be carried out in the file to determine if this rule has been violated. For example, the number of faces of the proper solid identified using the STL must be the even number. In addition, the number of edges must be divisible by three and the equation must be followed:

$$\frac{\text{No. of Faces}}{\text{No. of Edges}} = \frac{3}{2} \tag{2.3}$$

Leaking STL files: As mentioned earlier, the STL files should describe fully enclosed surfaces that represent the solids generated by the originating CAD system. In other words, STL data files should build one or more multiple entities according to Euler's solids rule.

No. of faces – No. of edges + No. of vertices = 2 × No. of bodies

If this rule does not hold, the STL file is said to leak and the file slices will not represent the actual model. There may be too few or too many vectors for a particular piece of paper. Slicing software may add additional vectors to close the outline, or it may just ignore additional vectors. Small defects may be ignored in this way. Large leaks may lead to unacceptable final models. Leaks can be created by facets that cross each other in 3D space as shown in Figure 2.10. This can result from poorly

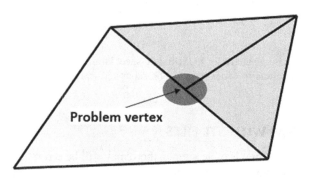

FIGURE 2.9 A case that violates the vertex-to-vertex rule.

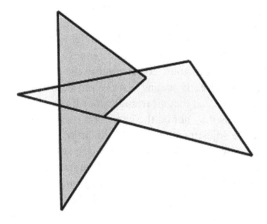

FIGURE 2.10 Two triangles intersecting each other in 3D space.

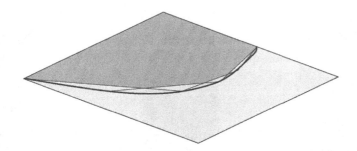

FIGURE 2.11   Two surface patches that do not match up with each other, resulting in holes.

generated CAD models, especially those that do not use Boolean operations to generate solids. A CAD model can also be created using a method that stitches surface patches together. If the triangular edges of the two surface patches do not suit each other, then the holes, like in Figure 2.11, may occur.

Degenerated facets: These facets are usually the result of a numerical truncation. The triangle may be so small that all three points are practically the same as each other.

After truncation, these points lay on top of each other, causing a triangle with no area. This can also occur when the truncated triangle does not return to any height and all three vertices of the triangle lie on a single straight line. While the resulting slicing algorithm will not cause incorrect slices, there may be some difficulty in checking algorithms and such triangles should actually be removed from the STL file. It is worth mentioning that while a few errors may occur in some STL files, most professional 3D CAD systems today produce high-quality, error-free results.

In the past, more common problems have arisen from surface modeling systems, which are now becoming scarcer, even in areas outside CAD engineering, such as computer graphics and 3D gaming software. Also, in earlier systems, STL generation was not properly tested and faults were not found in the CAD system. Potential problems are now well understood, and there are well-known algorithms for identifying and resolving these problems. However, the recent boom in home-use 3D printers has resulted in a wide range of software routines that are readily available but not thoroughly tested and may suffer from the problems mentioned in this chapter.

2.4   STL FILE MANIPULATION

Once a part has been converted to STL, there are only a few operations that can be performed. This is because the triangle definition does not allow for radical changes to the data. Associations between individual triangles are based solely on common points and vertices. A point or a vertex may be shifted, which will impact the connected triangles, but generating a regional effect on larger groups of points would be more difficult. Consider modeling a simple geometry, like the cut cylinder in Figure 2.12a. Making a minor change to one of the measurements may result in a very radical change in the distribution of the triangles.

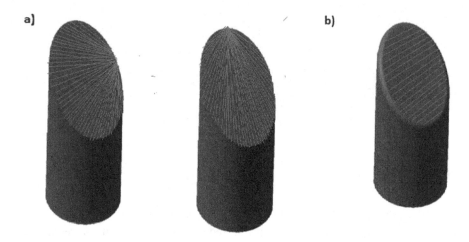

FIGURE 2.12   The STL files of a cut cylinder. Notice that while the two models in (a) are very similar, the position of the triangles is somewhat different. Adding a simple filet to (b) shows an even greater change in the STL file.

While it is possible to modify the model by reducing the number of triangles, it is quite easy to see that the boundaries in most models cannot be easily achieved. Adding the filet to Figure 2.12b shows an even more radical change to the STL file. In addition, if one were to attempt to move the oval representing the cut surface, the triangles representing the filet would no longer have a constant radius curve. Building models using AM are often done by people working in departments or companies that are different from the original designers. It may be that whoever builds a model may not have direct access to the original CAD data. Therefore, it may be necessary to modify the STL data before the part is to be built. The following sections explain the generally used STL tools.

2.4.1   VIEWERS

A variety of STL viewers is available, mostly as a free download. An example is the Marcam STL view (see Figure 2.13). Like many other systems, this software allows restricted access to the STL file, making it possible to view triangles, apply shading, display sections. By buying a complete version of the program, other tools may be used, for example, to allow the user to measure the part at various places, to annotate the part, to show sliced information, and to identify possible data problems. Often free tools allow for passive viewing of STL data, while more advanced tools allow for data modification, either by rewriting the STL or by providing additional information with STL data (such as measurement information). These viewers are often linked to part-building services and provided by the company as an incentive to use these services and to help decrease errors in data transfer, either from incomplete or inaccurate conversion of the STL or from incorrect interpretation of the design intention.

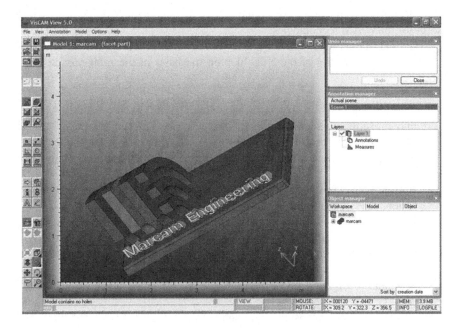

FIGURE 2.13   The VisCAM viewer from Marcam that can be used to inspect STL models.

2.4.2   STL MANIPULATION ON THE AM MACHINE

The STL data for a part consists of a set of points defined in space, based on an arbitrary point of origin. This point of origin will not be sufficient for the machine on which the component is to be designed. In addition, even if the part is correctly specified within the machine space, the user may choose to move the part to another location or create a duplicate to be installed next to the original component. Other tasks, such as scaling, changing orientation, and merging with other STL files, are all things that are routinely done using the STL manipulation tools on the AM machine.

The development of support structures is also something that would usually be expected to happen on the AM system. This would normally be done automatically and would be an operation applied to the downward-facing triangles. Supports would be enlarged to the base of the AM machine or to any upward-facing triangle placed directly below it. For some AM technologies, triangles that are only just moving away from the vertical (e.g. less than 10) may be ignored. Note, for example, the supports produced by the cup handle in Figure 2.6. There is little or no control over the placement of support or manipulation of STL model data with some AM operating systems. Looking at Figure 2.6 again, it may be possible to build the handle feature without so much support or even without support at all. A small amount of sagging around the handle may be visible, but the user would choose to have the model cleaned to remove the support content. If this type of control is required by the user, it may be necessary to purchase additional third party software such as MAGICS and 3-matic systems from Materialise. These third party tools can also be used to perform additional functions. For example, MAGICS has a range of modules that are useful for many AM technologies.

Other STL file manipulators may have similar modules as follows:

– Check the integrity of the STL files on the basis of the problems described above.

- Incorporating support structures, including tapered support features, which may make them easier to remove.
- Maximizing the use of AM computers, such as ensuring that the system is effectively packed with components, minimizing the number of support structures, etc.
- Adding serial numbers and markings to the parts to ensure correct identification, easy assembly, etc.
- Re-meshing of STL files that may have been generated using reverse engineering tools or other non-CAD-based programs. Such files may be overly large and can often be reduced in size without compromising the accuracy of the part.
- Segmenting large models or merging several STL files into one model data set.
- Performing Boolean tasks such as subtracting model data from a blank template insert tool to create a mold.

2.5 BEYOND THE STL FILE

The STL definition was created by 3D Systems right at the beginning of the development history of AM technology and has served the industry well. There are, however, other ways in which files can be defined for the creation of a slice. Furthermore, the fact that the STL file only represents surface geometry may cause problems for parts requiring certain heterogeneous content. Some of the issues surrounding this area will be discussed in this section.

2.5.1 DIRECT SLICING OF THE CAD MODEL

Since generation of STL files can be time-consuming and error-prone, there may be some advantage from using inbuilt CAD tools to directly generate slice data for AM machines. For most 3D solid modeling CAD systems, it is a trivial task to calculate the intersection of a plane with a model, thus extracting a slice.

This slice data would normally have to be processed to suit the AM drive system, but this can be handled in most CAD systems using macros. Support structures can be generated using standard geometry specifications and projected to the part from a virtual interpretation of the AM machine build platform. Although this approach has never been a popular method for generating sliced data, it has been studied as a research subject and even produced to suit a commercialized variant of the Stereolithography process by a German company called Fockle & Schwarz.

The main barrier to using this approach is that each CAD system must include a suite of different direct cutting algorithms for a variety of machines or technologies. This would be a cumbersome approach that may require periodic technology updates as new machines become available. There may be some future benefits in creating an integrated design and manufacturing solution, especially for niche applications or low-cost solutions. At present, however, it is more prudent to distinguish the creation of design tools from that of AM technology by using the STL format.

2.5.2   COLOR MODELS

Currently, many AM technologies are available in the market that can produce color variance components, including full color production, including 3D Systems color binder jetting technology, 3D Systems and Stratasys material jetting machines, and Mcor Technologies sheet lamination systems. Colored parts have proven to be very popular, and other color AM machines are likely to make their way to the market. The traditional STL file does not contain any details pertaining to the color of the component or any of its features. Coloring of STL files is possible and there are currently color definitions of STL files available, but you will be constrained by the fact that a single triangle can only be one particular color. It is therefore much easier to use the VRML painting options that allow you to allocate bitmap images to various facets. In this way, you can take advantage of the color options that the AM machine can give you.

2.5.3   MULTIPLE MATERIALS

From the previous section, color is one of the clearest examples of multiple material products that AM is capable of producing. As described in other chapters, parts can be made using AM from composite materials, with varying porosity rates or with regions containing discretely various materials. For many of these new AM technologies, STL is starting to become an obstacle. Because the STL description is for surface data only, it is therefore presumed that the solid material between these surfaces is homogeneous. This may not be the case, as we can see from the above. Although considerable consideration has been given to the issue of representations for heterogeneous solid modeling, there is still a great deal to consider until we can arrive at a standard to supersede STL for future AM technology.

2.5.4   USE OF STL FOR MACHINING

STL is used for applications beyond transforming CAD to additive manufacturing input. Reverse engineering packages can also be used to convert point cloud data directly to STL files without the need for CAD. Such technology, which is directly connected to AM, could conceivably form the basis for a 3D fax machine. Subtractive manufacturing is another technology that can easily make use of STL files. Subtractive manufacturing systems can easily use the surface data described in the STL to assess the machining boundaries. With some additional knowledge on

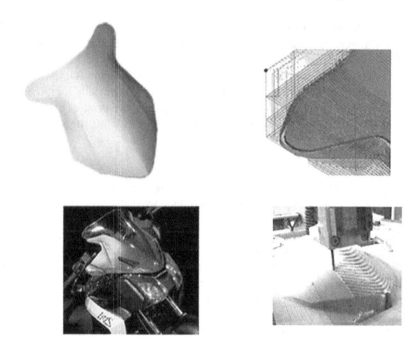

FIGURE 2.14   DeskProto software is used to extract machine tool paths from the STL file data to shape a mold to create a motorcycle windscreen.

the sizes of the starting block of material, tool, machining center, etc., it is possible to calculate machining strategies for the creation of a 3D surface model.

As stated earlier, it is likely that complex geometries cannot be completely machined due to undercutting features, internal features, etc., but there is no reason why STL files cannot be used to build computer-aided manufacturing (CAM) profiles for machining centers. For several years now, Delft Spline has been using STL files to create CAM profiles. Figure 2.14 shows the progress of the model to the tool used to manufacture the final product using their DeskProto software. The SRP (subtractive rapid prototyping) technique developed by Roland for their desktop milling machines is another technology that uses a combination of subtractive and additive processes to produce parts.

2.6   ADDITIONAL SOFTWARE TO ASSIST AM

As well as actually controlling the manufacturing process, other software systems can be helpful in running an efficient and effective AM-based facility.

2.6.1   SURVEY OF SOFTWARE FUNCTIONS

Such software may include one or more of the following functions.

Simulation: Many operating systems can perform a simulation of the machines operations in a build process, showing how the layers will be formed step by step in accelerated time. This can allow the user to detect obvious errors in the slice files and determine whether critical features can be built. This may be particularly important for processes like material extrusion where the hatch patterns can have a critical effect on thin wall features, for example. Some work has been done to simulate AM systems in order to get a better impression of the final result, including the images rendered to give an understanding of the surface roughness of the given layer thickness.

Build-time estimation: AM is an extremely automated process and the recent machines are very reliable and can operate unattended for a long period of time. It is very important to know when the construction will be completed for effective process planning. Knowing this will help determine when operators will be required to change their jobs.

Good estimates would also help to balance the builds; adding or subtracting a part from the work batch can ensure that the machine cycle is completed within a day-shift span, for example, making it possible to keep machines running at night unattended. Also, if you run multiple machines, it would be helpful to stagger builds throughout the shift to optimize the manual work required. Early build-time estimation program was highly inefficient, conducting a rolling estimate of the average build time per row. Since the time of the layer depends on the geometry of the part, such estimates could be very imprecise and vary wildly, especially at the beginning of the construction. Later software versions saw the benefit of more accurate build-time estimates.

Machine setup: Since every AM machine has an operating system that makes it possible to build, such systems can be very basic, especially when it comes to handling STL files. However, the determination of building parameters based on a given material is typically very detailed.

Monitoring: This is a relatively new feature of most AM systems. Although almost every AM machine will be connected either directly or indirectly to the Internet, it has traditionally been used to upload model files to the building site. Exporting information from the machine to the Internet or within the intranet was not common except for larger, more expensive machines. The easiest monitoring systems would provide basic information on the status of the build and how much longer it is to be completed. However, more complex systems may tell you how much material remains, the current status parameter such as temperatures, laser power, etc., and whether there is a need for manual intervention through the alerting system. Some monitoring systems may also provide video feedback on the build.

Planning: Simulation of the AM process running on a different computer may be helpful for those working in process planning. System planners can be able to decide what a building could look like, allowing for the possibility of preparing new jobs, evaluating uncertainty, or quoting. Again, such applications can be available from the AM system provider or from a third party vendor. The downside of third party applications is that it is more likely to be updated to meet the exact specifications of the customer.

2.6.2 AM PROCESS SIMULATIONS USING FINITE ELEMENT ANALYSIS

Finite element analysis (FEA) techniques are increasingly popular tools for predicting how the output of different manufacturing processes changes with changing process parameters, geometry, and/or material. Commercial software packages such as SYSWELD, COMSOL, Moldflow, ANSYS, and DEFORM are used to predict the results of welding, forming, molding, casting, and other processes.

AM technologies are particularly difficult to simulate using finite element predictive tools. For example, the multi-scale nature of metal powder bed fusion approaches such as metal laser sintering and electron beam melting are incredibly time consuming to accurately simulate the use of FEA-based physics. AM processes are inherently multi-scale in nature, and fine-scale finite element meshes that are 10 µm or smaller in size are needed to accurately capture solidification physics around the melt stream, whereas the total component size could be 10,000 times larger than the element size.

In three dimensions, this means that if we apply a uniform 10 µm mesh size, we would need 10^8 elements in the first layer and more than 10^{12} elements in total to capture the physics for a single part that fills much of a powder bed. Since rapid movement of a point heat source is used to create parts, capturing the physics requires a time step of 10 ms or less during laser/electron beam melting, which for a complete build would require more than 10^{10} total time steps. To solve this problem, it would take billions of years for a number of elements to take this manytime phases on a fairly high-speed supercomputer. Thus, to date all AM simulation tools are limited to predictions of only a small fraction of a part, or very small, simplified geometries. Using existing FEA tools, several researchers are looking for ways to make assumptions whereby they can cut and paste solutions from simplified geometries to form a solution for large, complex geometries.

This approach has the advantage of faster solution time; however, for large, complex geometries, these types of predictions do not accurately capture the effects of changing scan patterns, complex accumulation of residual stress, and localized thermal characteristics. In addition, minor changes to the input conditions may invalidate simplified solutions. Thus, a simulation infrastructure that can quickly build up a "new" response to any arbitrary geometry, input condition, and scan pattern is the ultimate goal of a predictive AM simulation tool. Researchers have recently begun using dynamic, multi-scale moving mesh to accelerate FEA analysis for AM. These types of multi-scale simulations are many orders of magnitude faster than standard FEA simulations. However, multiscale simulations alone are still too slow to allow complete part simulation, even on the world's fastest supercomputers. A new computational approach for AM is therefore needed, which can extend the FEA beyond its historical capabilities.

3DSIM, a new software start-up, is trying to do that for the AM industry. The software tools of 3DSIM include:

1. A new approach to formulating and resolving multi-scale moving mesh
2. A novel, finite element-based autosolver that predicts thermal evolution and residual stress very quickly in low thermal gradient regions
3. An insignificant number truncation Cholesky module, which eliminates the "null multiplication" calculations that occur when the finite element matrices are fixed, and

4. A stand-alone approach to the identification of periodicity in AM computations to enable feed-forward "insertion" of solutions in regions where periodicity is present and the solution is already known from a previous stage of time. These approaches are reported to reduce the time to solve large-scale AM problems by orders of magnitude. Taken together, these tools should make the full part of the problem solvable on a desktop GPU-based supercomputer in less than a day when algorithms are fully implemented in a combined software infrastructure. If realized, the ability to predict how process parameters and material changes affect part accuracy, distortion, residual stress, microstructure, and properties would be a significant advance for the AM industry.

2.7 THE ADDITIVE MANUFACTURING FILE FORMAT

It has already been mentioned that, although successful, there are numerous difficulties with regard to the STL format. As AM technologies move forward to include multiple materials, lattice frameworks, and textured surfaces, an alternative format is likely to be needed. In May 2011, the ASTM Committee F42 on Additive Manufacturing Technology published ASTM 2915-12 AMF Standard Specification for AMF Format 1.1. This file format is still under development, but has already been incorporated in some commercial and beta-stage applications.

Substantially more complex than the STL format, AMF aims to embrace a whole host of new parts descriptions that have hindered the development of current AM technologies. These shall include the following features:

Curved triangles: In STL, the normal surface lies on the same plane as the vertical triangle, but in AMF, the starting position of the normal vector does not have to lie on the same plane. If this is the case, the corresponding triangle must be curved. The definition of curvature is such that all triangle edges at that vertex are curved so that they are perpendicular to the normal and in the plane defined by the normal and the original straight edge (i.e. if the original triangle has straight edges rather than curved edges). By specifying the triangles in this way, far fewer triangles need to be used for a typical CAD model. This addresses the problems related with large STL files resulting from complex geometry models for high resolution systems. The curved approach of the triangle is still an approximation, since the degree of curvature cannot be too high. However, overall accuracy is significantly improved in terms of the deviation of the cusp height.

Color: Color can be nested in such a way that the main body of the part can be colored according to the function of the original design. Coloring of red, green, and blue may be applied along with a transparency value to vertices, triangles, volumes, objects, or materials. Note that many AM processes, such as vat photopolymerization, can make clear parts so that the transparency value can be an effective parameter. Color values may work along with other material-based parameters to provide a versatile way to control the AM process.

Texture: The color assignment described above cannot deal directly with the image data assigned to the objects. However, this can be achieved by the texture operator. Texture is allocated geometrically first, by scaling it to an element in such a way that individual pixels are added to the object in a uniform manner. These pixels will have the intensity, which is then assigned to the color. It should be noted that this is an image texturing process similar to computer graphics rather than a physical texture, such as ridges or dimples.

Material: It is possible to delegate specific volumes to be produced using different materials. Currently, Stratasys Connex machines and some other extrusion-based systems have the capacity to build multiple material parts. At the moment, the design of the parts requires a time-consuming redefinition process within the operating system of the machine. By having a material definition within AMF, it is possible to carry it all the way from the design stage.

Material variants: AMF operators may be used to changing the basic structure of the component to be produced. For example, a variety of medical and aerospace applications may require a lattice or porous structure. The operator can be used in such a way that the specified volume can be constructed using an internal lattice structure or porous material.

In addition, some AM technologies would be able to make parts of materials that gradually blend with others. A periodic operator can be applied to a surface that converts it into a physical texture, rather than the color mapping mentioned above. It is even possible to use a random operator to provide the AM part with unusual effects. It should be noted that a component designed and encoded using the AMF would almost certainly look different when it is constructed using different machines. This will be especially true for parts that are coded according to different materials, colors, and textures. Each machine will have the capacity to accept and interpret the design of the AMF according to its functionality.

For example, if a part is defined with a fine texture, a lower resolution process will not be able to apply it as well. Opaque materials will not be able to make great use of the transparency function. Some machines will not be able to make parts with multiple materials, and so on. It should also be noted that all machines should be able to accept the definition of geometry and make a part of the AMF defined part. As a result, AMF is backward compatible so that it can recognize a simple STL file, but with the ability to specify any conceivable design in the future.

2.8 SOLVED EXAMPLE

Part Build Time in STL

Time to complete a single layer:

$$Ti = \frac{Ai}{vD} + Td \qquad (2.4)$$

where Ti = time to complete layer i;
A_i = area of layer i;
v = average scanning speed of the laser beam at the surface;
D = diameter of the "spot size," assumed circular; and
T_d = delay time between layers to reposition the worktable

Once the Ti values have been determined for all layers, then the build cycle time is:

$$T_c = \sum_{i=1}^{ni} T_i$$

where Tc = STL build cycle time; and
nl = number of layers used to approximate the part

- Time to build a part ranges from one hour for small parts of simple geometry up to several dozen hours for complex parts

Example: A prototype of a tube with a square cross-section is to be fabricated using stereolithography. The outside dimension of the square = 120 mm and the inside dimension = 110 mm (wall thickness = 5 mm except at corners). The height of the tube (z-direction) = 90 mm. Layer thickness = 0.15 mm. The diameter of the laser beam ("spot size") = 0.30 mm, and the beam is moved across the surface of the photopolymer at a velocity of 600 mm/s. Compute an estimate for the time required to build the part, if 15 s are lost each layer to lower the height of the platform that holds the part. Neglect the time for post-curing.

Solution: Layer area A_i is the same for all layers.
$A_i = 120^2 - 110^2 = 2300$ mm^2.
Time to complete one layer T_i is the same for all layers.

$$Ti = \frac{Ai}{vD} + Td$$

$$Ti = \frac{2300}{600 \times 0.30} + 15$$

$$T_i = 27.7 \text{ s}$$

Number of layers

$$nl = \frac{90}{0.15} = 600 \text{ Layes}$$

$$T_c = 600(27.7) = 16620 \text{ s} = 277.0 \text{ min} = 4.61 \text{ hr}$$

2.9   EXERCISES

1. How would you change Flowchart 2.2 to include multiple contours?
2. Under what conditions would you like to combine more than one STL file together?

3. Write the ASCII STL file for a perfect cube, aligned with the Cartesian coordinate frame, starting at (0, 0, 0) and all dimensions are positive. Model the same cube on a CAD system. Does that render the same STL file? What happens when you make slight changes to the design of the CAD?
4. Why is it possible that a part might inadvertently be made 25 times too small or too large in any one direction?
5. Is it all right to ignore the vertex of the triangle that lies directly on the intersecting cutting plane?
6. Prove with a few simple examples that the number of faces divided by the number of edges is 2/3.
7. A prototype of a tube with a square cross-section is to be fabricated using stereolithography. The outside dimension of the square = 110 mm and the inside dimension = 100 mm (wall thickness = 5 mm except at corners). The height of the tube (z-direction) = 100 mm. Layer thickness = 0.10 mm. The diameter of the laser beam ("spot size") = 0.27 mm, and the beam is moved across the surface of the photopolymer at a velocity of 550 mm/s. Compute an estimate for the time required to build the part, if 10 s are lost each layer to lower the height of the platform that holds the part. Neglect the time for post-curing.

2.10 MULTIPLE-CHOICE QUESTIONS

1. The first step in each product design process is to _____ and conceptualize the function and appearance of the product.
    a) Summarize
    b) Imagine
    c) Visualize
    d) None of the above
    Ans: (c)

2. The STL format of the 3D CAD model captures all surfaces of the 3D model by stitching _____ of different sizes on its surfaces.
    a) Angles
    b) Triangles
    c) Planes
    d) All of the above
    Ans: (b)

3. The outside of the triangle is better established by the
    a) Left-hand rule approach
    b) Right-hand rule approach
    c) Thumb rule approach
    d) None of the above
    Ans: (b)

4. The appropriate value of the required offset would be largely dependent on the _____ of the AM process to be used.
   a) Resolution or accuracy
   b) Precision or accuracy
   c) Threshold
   d) All of the above
   Ans: (a)

5. The identification of the _____ to be assisted can be made by evaluating the angle of the standard triangle.
   a) Surface
   b) Plane
   c) Face
   d) Region
   Ans: (d)

6. Which technology requires a filling pattern to be produced within the enclosed boundary?
   a) Binder jetting
   b) Material extrusion
   c) Vat polymerization
   d) None of the above
   Ans: (b)

7. All geometric solid modeling CAD systems can generate _____ files from a valid, fully enclosed solid model.
   a) HTML
   b) DHTML
   c) STL
   d) All of the above
   Ans: (c)

3 Liquid-Based Additive Manufacturing Systems

3.1 3D SYSTEMS STEREOLITHOGRAPHY APPARATUS (SLA)

The stereolithography apparatus (SLA) process was the first commercialized RP process and represents the stereolithography process. Stereolithography, patented in 1986, started a rapid prototyping revolution. It works on the concept of solidifying a photosensitive resin using a UV light layer-by-layer laser to create a 3D model. Stereolithography uses photocurable resins that can be classified as epoxy, vinyl, or acrylate. Acrylics cure only about 75% or 80% since curing stops as soon as UV light is removed. Epoxies tend to heal even when the laser is not in contact. The device, as shown in Figure 3.1, consists of a platform which is moved down as each layer is formed in a resin-containing tank. In the X–Y plane, the laser light is moved by a positioning system. In some cases, a support structure must be set up to support the overhanging parts. Some of the commercial SLA machines are shown in Figure 3.2.

The stereolithography process transforms 3D computer image data into a series of very thin cross-sections, as if the object had been sliced into hundreds or thousands of layers. A vat of photosensitive resin comprises a vertically moving platform. The part under development is supported by a platform that moves down a layer thickness (usually about 0.1 mm or 0.004 in.) for each layer. The laser beam then traces a single layer of liquid polymer onto the surface of the vat as shown in Figure 3.3. The ultraviolet light causes the polymer to harden exactly at the point where the light reaches the surface. As shown in Figure 3.3a, the model is constructed on a platform located just below the surface of a vat of liquid epoxy or acrylate resin.

A low-power, highly concentrated UV laser traces the first layer (see Figure 3.3b as the laser traces in the center of the layer), solidifying the cross-section of the model while leaving excess areas liquid. The UV laser is controlled by a galvanometer scanner to generate X–Y motion, which means that the table does not need to move in the x and y directions. Then, the elevator progressively lowers the platform to the liquid polymer as shown in Figure 3.3c. The laser tracks down from the left. The weeper coats the solidified layer with the liquid, and the laser traces the second layer above the first layer. The process is repeated until the prototype is completed (Figures 3.3d and 3.3e). Afterward, the solid part is removed from the vat and the excess liquid is rinsed clean as shown in Figure 3.3f. In all cases where a part is built, there is a small structure attached to the bottom called "supports." Its purpose is to raise the part off the platform and provide a "bridge" type structure that only touches the part by small points. This structure is removed after the part is finished. The brackets are broken off and the model is then put in an ultraviolet oven for

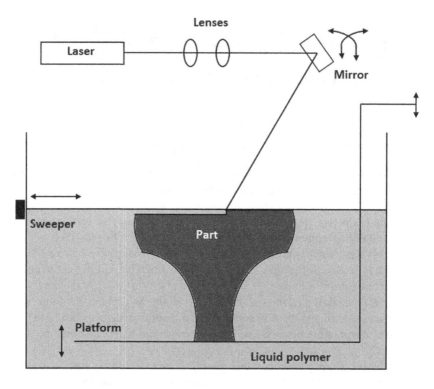

FIGURE 3.1   An illustration of the SLA process.

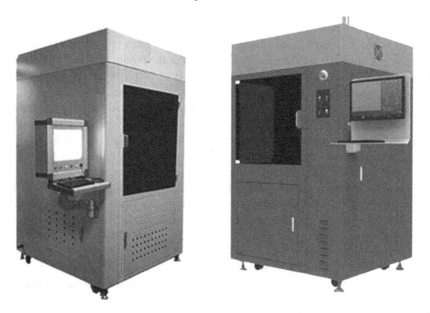

FIGURE 3.2   Commercial SLA machines.

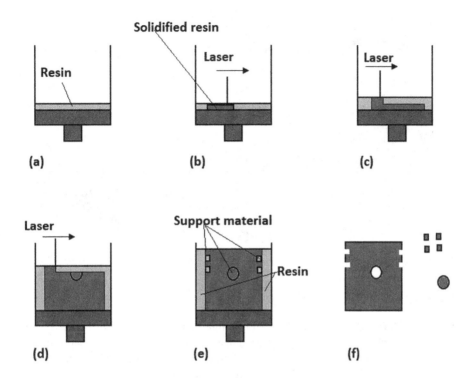

FIGURE 3.3 Stereolithography process step-by-step: (a) A layer of resin to be solidified on a platform; (b) UV laser selectively traced out the first layer; (c) Second layer with laser tracing from the left; (d, e) Repeat to build the remaining of the layers; (f) The final part after the support structures are removed.

complete curing. Current stereolithography processes have advanced their technology. For example, the Viper Pro SLA system by 3D Systems has adjustable beam sizes to accelerate part building speed. It has the capacity to build a volume of 1500 × 705 × 500 mm³.

An example of a part of a dashboard is shown in Figure 3.4. For example, in Formula 1, Renault uses only SLA and selective laser sintering (SLS) models to test the wind tunnel. There are many good day-to-day examples that people can easily associate with the benefits of today's and the future development of technology.

The uniqueness of this process is its resolution and precision. The end product is a very near physical model or prototype with 3D design—giving designers, engineers, suppliers, sales managers, marketing managers, and prospective customers the ability to experience a new product or prototype. In this way, concept improvements can be created quickly and cheaply, ensuring customers of the best possible product as quickly as possible.

It has become a much-used technology in so many industries. Examples include aircraft, arms, automotive, consumer electronics, consumer products, toys, construction equipment, medical equipment, surgical applications, and dental applications.

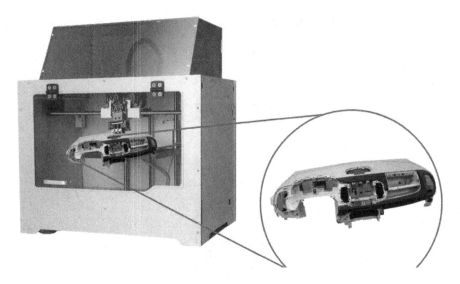

FIGURE 3.4   An SLA model of a dashboard part.

Since this was the first technology, stereolithography is assumed to be the bench-mark by which other technologies are judged. Early stereolithography prototypes were relatively fragile and prone to curing-induced warping and distortion, but the latest advancements have largely corrected these problems.

It is essential that the RP process is very stable, and stereolithography is a pro-cess that has the character of an unattended building process—once it is started, the process is fully automatic and can be unattended until the process is completed. It has good dimensional accuracy, too. The process is capable of maintaining the dimensional accuracy of the components to within ±0.1 mm. Due to the liquid prop-erties, the produced product has a good surface finish—glass-like finishing can be achieved on the top surfaces of the part. While stairs can be found on the side walls and curved surfaces between the construction layers, 3D Systems Inc. has devel-oped Quickcast software for building parts with a hollow interior which can be used directly as an investment casting wax pattern. This is a perfect example of how RP processes are applied to tooling applications.

One drawback is that the absorption of water into the resin over time in thin areas will result in curling and warping. The cost of the system is relatively high and the material available is only photosensitive. The parts cannot often be used for durabil-ity and thermal testing. In most cases, the parts were not completely healed by the laser inside the vat. This is due to the fact that, when a laser is curing a spot, the energy is a cone shape, as shown in Figure 3.5, and during processing, there are some uncured regions throughout the part, and therefore a post-curing process is normally required. The price of the resin and laser guns is very high. In addition, the optical sensor requires regular fine-tuning in order to maintain its optimal operating condi-tion, which will be very costly, as are the labor requirements for post-processing, in particular cleaning.

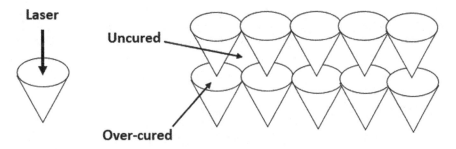

FIGURE 3.5   The cone features generated by the laser curing process resulting in uncured regions throughout the part.

3.2   STRATASYS POLYJET

There is an alphabet soup of AM methods that have been developed over the last 30-plus years. The big players are 3D Systems and Stratasys, which have a combined capitalization of $4.15 billion and have been beating competitors and service offices and accumulating patent portfolios at a massive rate. Unfortunately for them, their stock values have fallen just as quickly recently. The industry is cautiously anticipating Hewlett-Packard's entry. Hewlett-Packard has a new technology in 2016 that is said to be more capable.

FDM: Fused deposition modeling was patented in 1989 by Stratasys. Stratasys bought Makerbot for over $400 million in 2013. This technology has become quite famous among hobbyists as patents have expired, and the gold rush of industries and individuals has started to make low-cost printers that produce components from ABS (acrylonitrile butadiene styrene, a traditional thermoplastic polymer—Lego is made from the same material), PLA (polylactide, a biodegradable thermoplastic and aliphatic polyester derived from renewable resources).

FFF: Fused filament manufacturing is equivalent to FDM; however, the term is unrestricted by Stratasys, a trademark of FDM.

SLA: Stereolithography was patented in 1986 by 3D Systems. This technique uses photopolymers exposed to UV lighter lasers to harden tiny liquid goo elements that, when aggregated, create a solid surface. SLS/DMLS/SLM: Selective laser sintering/direct metal laser sintering/selective laser melting is a method that uses focused lasers to melt powders (plastic or metal) into tiny pools of material that are then cooled and aggregated into pieces. The fight for patent rights can continue until all of them have expired.

CJP: ColorJet printing was developed by MIT in 1993 and sold by ZCorp until it was purchased by 3D Systems in 2012. In this process, a layer of powder is infused with a liquid binder and cured to form the component. This method is remarkable because it works like a color inkjet printer.

PolyJet: Invented by Objet Geometries in 1998, Stratasys acquired PolyJet in 2011. It is a 3D-printing system that uses two or more photopolymer resins stored in tiny droplets, such as an inkjet printer, which are mixed in real time and cured with

UV light to create a solid object. PolyJet technology can produce over 100 types of durable plastic materials, including hard, soft, clear, and full color. PolyJet is a strong 3D printing technology that creates smooth, accurate parts, prototypes, and tools, with microstructural layer resolution and accuracy down to 0.014 mm, thin walls, etc. Complex geometries can be developed using the widest range of materials available with any technology.

Benefits of PolyJet:

i. Develop smooth, accurate prototypes that convey esthetics to the final product
ii. Produce accurate molds, jigs, fixtures, and other manufacturing devices
iii. Achieve complex shapes, complex details, and delicate features
iv. Implement the vast variety of colors and materials in a single component for unmatched efficiency

3.2.1 FUSED DEPOSITION MODELING FROM STRATASYS

Fused deposition modeling (FDM) produced and developed by Stratasys, United States, is by far the most common extrusion-based AM technology. FDM uses a liquefied polymer heating chamber, which is fed into the system as a filament. The filament is pushed into the chamber by the arrangement of the tractor wheel, and it is this thrust that generates the pressure of the extrusion. A typical FDM machine with a picture of the extrusion head can be seen in Figure 3.6.

FIGURE 3.6 Typical Stratasys machine showing the outside and the extrusion head inside.

The original FDM patent was issued to Stratasys founder Scott Crump in 1992 and the business has gone from strength to strength to the point that there are more FDM machines than any other AM machine in the world. The key strength of FDM lies in the variety of materials and the efficient mechanical properties of the resulting components developed using this technology. Parts made using FDM are among the best parts for any polymer-based additive manufacturing process. The biggest downside to using this technology is the speed of building. As stated earlier, the inertia of the plotting heads means that the maximum speeds and accelerations that can be reached are much lower than the other systems. In addition, FDM needs material to be plotted in a point-wise, vector-mode that involves a number of directional changes.

3.2.2 MATERIAL JETTING MACHINES

The major three companies involved in the growth of the RP printing industry are still the key players offering printing machines: Solidscape, 3D Systems, and Stratasys (after their merger with Objet Geometries). Solidscape sells the T66 and T612, both descendants of the previous ModelMaker line and based on the first generation melted wax technique. Each of these devices uses two single jets—one for depositing a thermoplastic part of the material and one for depositing a waxy support material—to shape 0.0005 inch thick layers. It should be observed that these machines often fly-cut layers after deposition in order to ensure that the layer is flat for the next layer. Due to the slow and precise construction style as well as the waxy materials, these tools are often used for the production of investment castings for the jewelry and dentistry industries. 3D Systems and Stratasys offer machines with the ability to print and process acrylic photopolymers. Stratasys markets the printer series Eden, Alaris, and Connex. These machines print a variety of different acrylic-based photopolymers materials in 0.0006 inch head layers containing 1,536 separate nozzles, resulting in fast, linear deposition performance, as compared to the slower, point-wise approach used by Solidscape. Each photopolymer layer is immediately cured by ultraviolet light as it is printed, producing fully cured models without post-curing. Support structures are made of gel-like material, which is extracted by hand and water jetting, see Figure 3.7 for the example of the Stratasys PolyJet method, which is used in all Eden machines. The Connex line of machines has multi-material functionality. For several years, only two different photopolymers could be printed at one time; however, by automatically changing the construction types, the computer can print up to 25 different effective materials by varying the relative composition of the two photopolymers. Machines are emerging that print a growing number of materials.

In comparison with Stratasys, 3D Systems markets ProJet printers with printing layers 0.0016 inch thick use of heads with hundreds of nozzles, half for part material and half for support material. Layers are then flashed with ultraviolet light, which triggers the polymerization induced by the photo. The ProJets are the third generation of the 3D Systems multi-jet modeling family, following the ThermoJet described above and the InVision series.

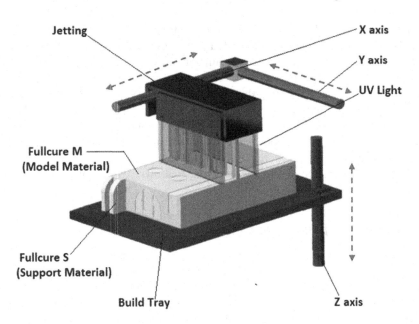

FIGURE 3.7   Stratasys PolyJet build process.

3.3   3D SYSTEMS' MULTI-JET PRINTING SYSTEM (MJP)

MJP or multi-jet printing is an inkjet printing process that uses piezo printhead technology to either deposit photocurable plastic resins or cast wax materials layer by layer. MJP is used to build parts, patterns, and molds with fine detail to discuss a wide range of applications. These high-resolution printers are economical to own and operate and use a separate, malleable, or dissolvable support material to make post-processing a breeze. Another major benefit is that the removal of support material is practically hands-free and allows even the most delicate features and complex internal cavities to be rigorously cleaned without damage (Figure 3.8).

MJP printers offer the highest resolution Z-directional resolution with layer thicknesses of up to 16 microns. In addition, selectable print modes allow the user to choose the perfect combination of resolution and print speed, making it easy to find a combination that suits your needs. Parts have a smooth finish and can achieve SLA-competing precision for many applications. Recent advancements in materials have enhanced the durability of plastic materials and are now suitable for some end-use applications. One good thing about MJP printers is that they are office compatible, using regular office electricity to provide easy and inexpensive access to high-quality prototypes and indirect development aids. Office compatibility with their capabilities makes MJP printers ideal for direct investment casting applications in jewelry, dental, medical, and aerospace applications where digital workflows deliver significant time, labor, quality, and cost advantages. MJP wax printers also provide a digital drop-in alternative to conventional waste-wax casting processes, reducing time-consuming and expensive process steps while utilizing traditional casting

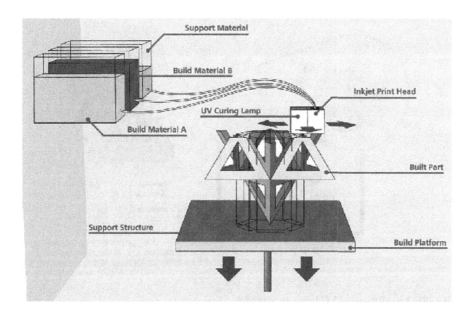

FIGURE 3.8 Multi-jet printing system (MJP).

methods and materials. All in all, these printers can print virtually any geometry and deliver scalable, high-volume throughput.

3.4 ENVISIONTEC'S PERFACTORY

Many companies are selling VP systems focused on mask projection technology, including EnvisionTEC and 3D Systems. New companies in Europe and Asia have also recently launched the market for MPVP (mask projection vat photopolymerization) systems. EnvisionTEC first introduced its MPVP systems in 2003. They now have a range of machine lines with different build envelopes and resolutions depending on the MPVP process, including Perfactory, Perfactory Desktop, Aureus, Xede/Xtreme, and Ultra. Variants of some of these models are available, including specialized Perfactory machines for dental restoration or hearing aid shells. A photograph of the Perfactory Standard machine is shown in Figure 3.9 and its technical specifications are shown in Table 3.1.

Their machines are very similar, schematically, to the Georgia Tech machine in Figure 3.10 and use a lamp to illuminate the DMD and the vat. However, several of their machine models have an essential difference: They build upside-down parts and do not use a coating mechanism. The vat is illuminated upward vertically through a clear window. After the system radiates a layer, the cured resin sticks to the window and cures to the previous layer. The build platform pulls out of the window at a slight angle to gently segregate the part from the window. The benefit of this approach is threefold. First, there is no need for a separate coating mechanism since gravity forces the resin to fill the area between the cured part and the window. Second, the

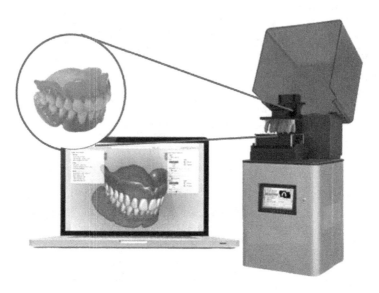

FIGURE 3.9   EnvisionTEC Perfactory model.

TABLE 3.1
Specifications on EnvisionTEC Perfactory Standard Zoom Machine

Lens system		f = 25–45 mm
Build envelope	Standard	190 × 140 × 230 mm
	High-resolution	120 × 90 × 230 mm
Pixel size	Standard	86–163 µm
	High-resolution	43–68 µm
Layer thickness	25–150 mm	

top surface of the vat being irradiated is a flat window, not a free surface, which enables more precise layers to be produced. Third, they invented a construction process that would remove a daily vat. Instead, they have a supply-on-demand feed system. The disadvantage is that small or fine features can be damaged when the cured layer is separated from the window. In 2008, 3D Systems introduced their V-Flash machine. It uses MPVP technology and a novel approach to material handling. The V-Flash was designed to be an affordable prototyping machine (under $10,000) that was as easy to use as a typical home inkjet printer. Its built-in envelope was 230 × 170 × 200 mm (9 × 7 × 8 in.). Parts were built upside-down during service. For each layer, the blade coated a layer of resin onto a film that spanned the building chamber. The construct platform slid down until the resin layer and the film was contacted by the platform or the in-process component. The cartridge provided for each layer a supply of unused film. That layer was cured by the UV Imager machine, which consisted of the MPVP technology. Some rinsing of the part was required, similar to

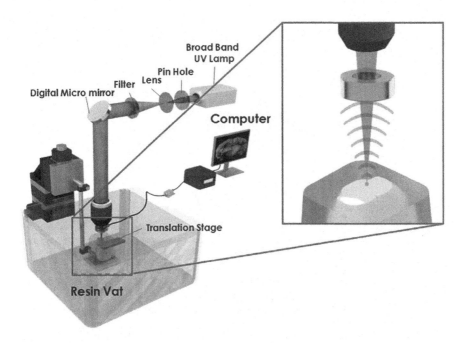

FIGURE 3.10 Schematic and photo of mask projection VP machine.

SL, and support structures may have to be removed during the post-processing phase of part manufacturing.

3.5 CMET'S SOLID OBJECT ULTRAVIOLET-LASER PRINTER (SOUP)

A number of Japanese companies focused on AM technology in the early 1980s and 1990s. This included start-up companies such as Autostrade (which appears to be no longer in operation). Large companies such as Sony and Kira, which have established subsidiaries to build AM technology, have also been involved. Much of Japanese technology was based on the process of photopolymer curing. With 3D Systems prevalent in much of the rest of the world, these Japanese companies have struggled to find a market, and many of them have failed to become commercially viable, even though their technology has shown some initial promise. Some of this may have resulted in the unusually slow adoption of CAD technology in the Japanese industry in general. While CMET still seems to be doing very well in Japan, you are likely to find more non-Japanese machines than home-grown ones in Japan. There is, however, some evidence that this is starting to change. Solid object ultraviolet-laser plotter (SOUP) is what Mitsubishi's CMET (Tokyo, Japan) calls its SLA-like stereolithography system.

CMET sold 56 units to organizations such as Mercedes, Fujitsu, Matsushita Electric, two Japanese universities, and Dornier Deutsche Aerospace in Germany. In addition to the consumer, Dornier is also a SOUP distributor in Europe. CMET sells variations of the SOUP 600 and 850 brands. The 600 uses either Helium Cadmium

SOLID OBJECT ULTRAVIOLET-LASER PRINTER

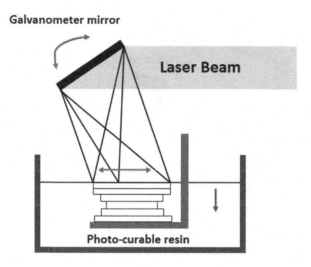

FIGURE 3.11   Schematic of SOUP system. (Adapted from CMET Brochure.)

or Argon ion laser and produces parts up to 600 × 400 × 400 millimeters (24 × 16 × 16 inches) of epoxy resin. The 850 uses the more powerful Argon ion laser and produces parts of up to 850 × 600 × 500 millimeters (33 × 16 × 20 inches). Instead of using a galvanometer mirror x–y scanner, the SOUP system uses an X, Y plotter mechanism to direct the laser light to the surface of the liquid resin. As a result, the laser beam remains perpendicular to the resin surface, minimizing unwanted light spread. Figure 3.11 shows the SUP System Schematic (adapted from the CMET Brochure).

3.6   ENVISIONTEC'S BIOPLOTTER

EnvisionTEC is the world's leading provider of professional 3D printers and materials. Founded in 2002 with its pioneering commercial DLP printing technology, EnvisionTEC is now selling 3D printers based on six distinct technologies that build objects from digital design files. The company's premium 3D printers serve a wide range of medical, technical, and industrial markets and are respected for accuracy, surface quality, flexibility, and speed.

Additive production is also one of the disruptive techniques in the area of biomaterials. In particular, the capacities provided by EnvisionTEC's 3D-Bioplotter allow us to build prototype implants with a special architecture that could not be created in any other way. Another example is the EnvisionTEC 3D-Bioplotter capable of producing a vast variety of soft and hard scaffolds using single or multiple materials (see Figure 3.12).

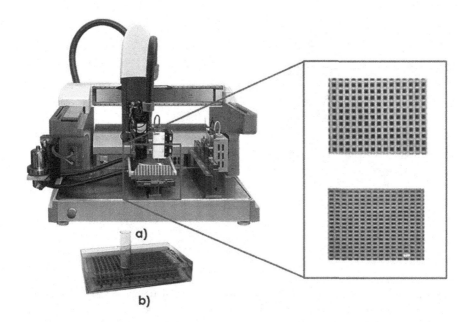

FIGURE 3.12 (a) Fourth generation 3D Bioplotter manufacturer series; (b) Schematic of the building platform. (EnvisionTEC 2017.)

Extrusion processes have been extensively used in the medical field, for example, to develop a more realistic airway trainer (Figure 3.13a), for the fabrication of customized molds for the pressing of a thin titanium sheet that will act as an orbital floor implant (Figure 3.13b), for the fabrication of bone bio-models for in-depth assessment and pre-surgical rehearsal resulting in a smoother operation process in which implants are more accurately fitted to the curvature of the patient's bone (Figure 3.13c), and the fabrication of exoskeletons (Figure 3.13d).

3.7 REGENHU'S 3D BIOPRINTING

The RegenHU 3DDiscovery is a 3D bioprinter developed by RegenHU, a manufacturer based in Switzerland. Bioprinting is a precise deposit of biomaterials such as cells, proteins, bacteria, and bio-gels in 2D or 3D. This technology can be used as a method for high-performance applications or to replicate biological systems that are similar and more precise to real living systems for study, testing, and diagnostic purposes. The aim of the RegenHU 3DDiscovery was to explore the potential of 3D tissue engineering through a bioprinting approach.

RegenHU provides state-of-the-art bioprinting solutions to enable your scientific and clinical ambitions. RegenHU 3D bioprinters are used for tissue engineering, personalized medicine, regenerative medicine, and basic drug discovery experiments. RegenHU's 3D bioprinting platforms are known worldwide for their industry-leading flexibility and precision. Customers use RegenHU 3D bioprinters to print bone, muscle, tendon, skin, kidney, liver, and lung tissue (Figure 3.14).

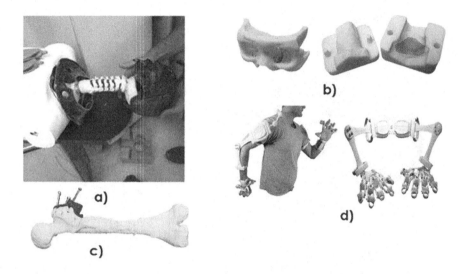

FIGURE 3.13 (a) Technician assembling a prototype airway trainer; (b) Patient's missing orbital floor (left) versus original shape before impact (right) and the customized mold for titanium sheet pressing; (c) Corrective osteotomy (realignment of bone from deformity) to complex bone fractures; (d) Custom orthopedic exoskeleton.

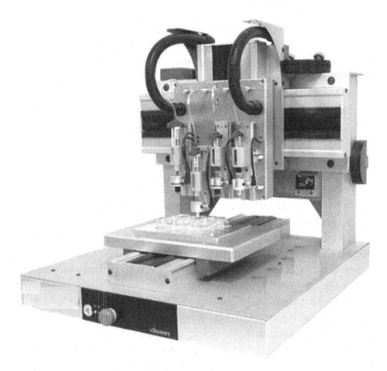

FIGURE 3.14   RegenHU's 3D bioprinter.

3.8   RAPID FREEZE PROTOTYPING

Rapid freeze prototyping (RFP) is a relatively new solid freeform manufacturing process that builds a three-dimensional part according to the CAD model by depositing and freezing water droplets layer by layer.

Rapid freeze prototyping is a solid freeform manufacturing (SFF) method that uses ice water as its medium. The system consists of a pressurized water container unit, an X–Y table for manipulating the plate to obtain the correct geometry of the component, a Z-axis elevator for raising the nozzle for successive layers, a circuit-driven nozzle, and a freezer. Figure 3.15 shows the schematic of the configuration. The nozzle is a precision micro-dispensing drop-on-demand nozzle that is cyclically opened by a function generator. The nozzle is supplied with water through a Teflon tube. Once the water leaves the Teflon tube, it enters the nozzle where it encounters stainless steel, polyphenylene sulfide, polyetherketone, ethylene/propylene rubber, butyl, epoxy, and finally sapphire before it is released to the substrate below. The materials with which the water comes into contact are critical because, if the water sticks to these materials, the flow rate would be greatly reduced due to adhesion. Materials that comprise the feed tube and the nozzle are materials that the water flows through with minimal adhesion.

There are many advantages to using rapid freeze prototyping over other SFF techniques, including stereolithography, fused deposition modeling, selective laser sintering, laminated object manufacturing, three-dimensional printing, and direct material deposition. Many SFF techniques use various materials in their methods, including UV curable resin, wax, ABS, metal/ceramic/polymer powders, and adhesive coated papers. As RFP uses water as a working material, the working environment is much

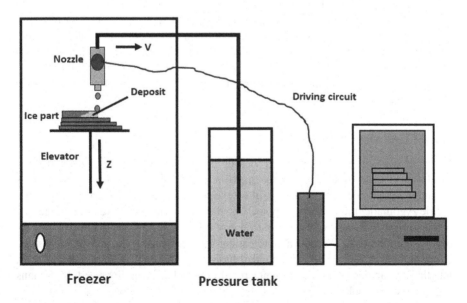

FIGURE 3.15   Principle of rapid freeze prototyping.

cleaner than alternative SFF techniques. Water is also easily available and inexpensive to use. RFP also provides quick construction time, less energy usage, clean and simple substrate detachment, and a very good surface finish.

3.9 FDM 3D PRINTING FOR ZYGOMATIC IMPLANT PLACEMENT MOCK SURGERY FOR PROSTHODONTIC DENTISTRY: A CASE STUDY

Dr. Priya Gupta, PG Student at VSPM Dental College, Nagpur, India, Dr. Saee Deshpande, VSPM Dental College, Nagpur, India, and Mr. Abhijeet Raut, Research Fellow, VNIT, Nagpur (MS), India, have done this case study of zyogomatic implant placement mock surgery for prosthodontic dentistry.

The quality of surgery depends on the anatomy of surgery. Three-dimensional (3D) anatomy visualization is very important for better understanding and a better outcome. Complex maxillofacial surgeries are usually time-consuming procedures due to the need for initial medical imaging for root complications, the determination of an appropriate course of treatment, pre-operative planning, before and after surgery. Pre-operative preparation is the most important and usually aims at executing an ideal procedure that minimizes the length of surgery and any possible problems that might occur in the future. For traditional dental procedures, the process begins by collecting the patient's impression using a powder mold until it is moved to the dental grade of alginate powder coated. These topography data are generally sufficient for the majority of cases; however, they are less acceptable for intricate cases where an implantable device may be essential.

In addition, medical imaging data offer little insight, since patient representation is a collection of two-dimensional planar images. In such situations, 3D information on the internal bone tissue structure is necessary for optimal surgical planning in order to determine the precise geometry of the affected area of interest and to properly assess the size and orientation of the implant. More recently, 3D printing technology has shown tremendous potential to increase and streamline pre-operative surgery preparation and care.

3.9.1 ZYGOMATIC IMPLANT

In recent years, edentulous patients seeking fixed recovery have increased. Yet the condition is unacceptable when the maxilla is heavily resorbed and atrophic. A zygomatic implant can be a successful way to rehabilitate a badly resorbed maxilla. While zygomatic implants are responsive to the technique, they have predictable outcomes if performed correctly. The purpose of this case study was to modify the facets of the implant and to create a more patient-specific design. The idea of developing two abutment designs with an expanded base plate and modified abutment widths, lengths, and angles to suit the patient's requirements has been discussed. Physicians and researchers took advantage of 3D printing technology and began a case study in 2018. The 3D printed prototype of the patient's jaw was designed using CBCT data for ease of preparation and comprehension. A mock surgery was performed in the model as shown in the figure. The proper location of the implant and the expected result of the operation were evaluated. The final optimized design was validated by a finite element analysis (FEA) to analyze the force distributions that are likely to be observed after surgery. This process is vitally important to ensure that there are no structural flaws in the design, which could otherwise lead to a catastrophic failure. It was relevant in the present case, due to the high force exerted on the maxilla and mandible elements during eating and chewing and the ultimate goal was to ensure patient health and a good outcome of treatment (Figures 3.16 and 3.17).

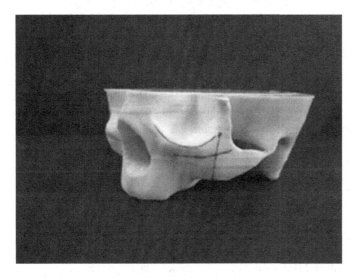

FIGURE 3.16 Shows the anatomical landmarks drawn in the model for the future position of zygomatic implants.

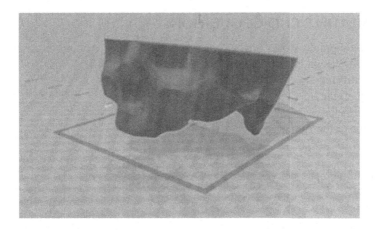

FIGURE 3.17 CAD model.

3.9.2 CONCLUSION

In this case study, Dr. Priya Gupta, Dr. Saee Deshpande, and Mr. Abhijeet Raut, along with the authors of the book, demonstrated the synergistic use of CT medical imaging, 3D CAD, and additive manufacturing in order to carry out the process of optimizing surgery planning and designing a personalized, patient-specific 3D model. And the 3D printing technology makes a complex process simpler for doctors. The use of medical imaging to establish a representative patient model has allowed the design of the zygomatic implant to be based on the specific anatomical characteristics of the patient. Using low-cost FDM 3D printing can also create medical models that allow mock surgery to be performed to test and optimize the design along with streamlining the procedure for actual surgery. Pre-surgical preparation has made the low-cost model possible and reduces the difficulty and time spent on the surgical innovation. This was also instrumental in achieving superior end esthetics and functionality. The findings of this case study provide a basis for future 3D printing of implantable devices. Such methodologies can result in lower costs for healthcare providers, thus increasing surgical effectiveness and enhancing the quality of treatment provided to the patient.

3.10 EXERCISES

1. Describe the concept of 3D Systems' stereolithography apparatus (SLA).
2. Discuss Stratasys PolyJet with its advantages.
3. Explain fused deposition modeling from Stratasy.
4. Explain 3D Systems' multi-jet printing system (MJP).
5. Discuss EnvisionTEC's Perfactory.
6. Describe CMET's solid object ultraviolet-laser printer.
7. Describe EnvisionTEC's Bioplotter.
8. Discuss RegenHU's 3D bioprinting.
9. Explain rapid freeze prototyping.

3.11 MULTIPLE-CHOICE QUESTIONS

1. Stereolithography works on the concept of _____ a photosensitive resin using a UV light layer-by-layer laser to create a 3D model.
   a) Solidifying
   b) Liquefying
   c) Modifying
   d) None of the Above
   Ans: (a)

2. Stereolithography was patented by 3D Systems in
   a) 1984
   b) 1986
   c) 1966
   d) 1989
   Ans: (b)

3. Fused deposition modeling (FDM) was patented by Stratasys in
   a) 1984
   b) 1986
   c) 1966
   d) 1989
   Ans: (d)

4. EnvisionTEC is the world's leading provider of professional 3D printers and materials and was founded in
   a) 2001
   b) 2002
   c) 2010
   d) 2003
   Ans: (b)

5. RegenHU 3DDiscovery is a 3D bioprinter developed by RegenHU, a manufacturer based in
   a) Germany
   b) Japan
   c) China
   d) Switzerland
   Ans: (d)

6. The aim of RegenHU 3DDiscovery was to explore the potential of _____ through a bioprinting approach.
   a) Bone engineering
   b) 3D tissue engineering
   c) Cell engineering
   d) None of the above
   Ans: (b)

7. Bioprinting is a precise deposit of _____ such as cells, proteins, bacteria, and bio-gels in 2D or 3D.
   a) Biocells
   b) Biochemicals
   c) Biomaterials
   d) All of the above
   Ans: (c)

8. Rapid freeze prototyping (RFP) is a relatively new solid freeform manufacturing process that builds a three-dimensional part according to the CAD model by depositing and _____ water droplets layer by layer.
   a) Freezing
   b) Solidifying
   c) Evaporating
   d) Liquefying
   Ans: (a)

4 Solid-Based Additive Manufacturing Systems

4.1 STRATASYS FUSED DEPOSITION MODELING (FDM)

Fused deposition modeling is a layer AM process that uses a thermoplastic filament by fused deposition. FDM was trademarked by Stratasys Inc. in the late 1980s and a similar term is FFF. The filament is extruded from a nozzle to print a cross-section of an item and then travels up vertically to replicate the procedure with a new layer (Figure 4.1).

The most frequently used FDM materials are ABS, PLA, and PC (polycarbonate), but new blends containing wood and stone as well as rubber-like filaments can be found. Compared to ABS, PLA reacts differently to moisture, to UV ageing with discoloration, and to material removal. In order to forecast the mechanical behavior of the FDM components, it is important to understand the material properties of the raw FDM process material and the effect that FDM builds parameters have on the properties of anisotropic materials (Ahn et al. 2002). The support material is often made from another material and is detachable or soluble from the actual material at the end of the manufacturing process (with the exception of low-cost solutions using the same raw material). The disadvantages are that the resolution on the z-axis is low compared to the other AM process (0.25 mm); therefore, if a smooth surface is needed, a finishing process is needed, and it is a slow process that sometimes takes days to build large, complex parts (Wong and Hernandez 2012).

FDM technology is the most popular 3D desktop printer and the cheapest professional printer to date. FDM was invented in the 1980s by Scott Crump (1992, 1994). The main strength of FDM lies in the variety of the materials and the efficient mechanical properties of the resulting components that have been produced using this technique. Components manufactured using FDM are among the best components for any polymer-based additive manufacturing process. The main drawback to the use of this technology is the speed of construction. The inertia of the plotting heads means that the maximum speeds and accelerations that can be achieved are slightly lower than other instruments. In addition, FDM allows information to be plotted in a point-wise, vector-mode that involves a number of directional changes.

4.2 SOLIDSCAPE'S BENCHTOP SYSTEM

Solidscape was founded in 1993 under the name of Sanders Prototype, Inc. by Royden C. Sanders to create PC-based 3D wax printers for rapid prototyping and master molds for investment casting. Sanders Prototype was initially located in

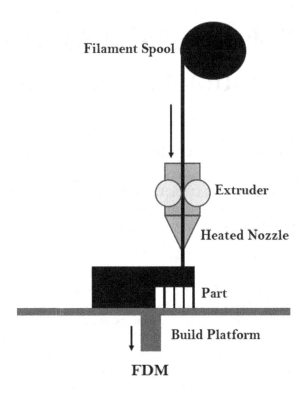

FDM

FIGURE 4.1 FDM.

Wilton, New Hampshire, and then moved to its current location in Merrimack, New Hampshire, United States. A new management team was set up in early 1998 and a major reorganization followed.

Sanders Prototype was renamed Solidscape Inc. in the fall of 2000. The first product was the Model Creator, a DOS-based desktop printer capable of producing high-resolution, three-dimensional wax models developed in CAD software packages. This machine was precise to less than 1,000th of an inch, allowing operators to produce very small, very detailed models. The wax models could then be cast without the need for a master pattern or a rubber mold. Solidscape's machines have established themselves as a favorite among custom jewelers, who enjoy the ability to create custom designs for customers and deliver finished goods faster and more accurately than by hand.

The first computer of Solidscape was Model 6 PRO. In relation to the vacuum cleaner, a desk-sized tower comprising an Intel 486DX processor was shipped on a standard motherboard, a 15-inch CRT monitor, and a keyboard. A proprietary interface card that interacted with the printer was also mounted onto the PC. The computer was running MS-DOS. The machine had to prepare the CAD models (converting them from the STL file to the proprietary format that the printer could use) and run the printer.

Conversion for most files took several hours to complete and several more to print. Depending on the model to be developed, the entire process from file to finished output frequently took 24–30 hours. Most of these units were development models, and very few were sold.

The 6 PROs were revised to become the Modelmaker in 1997.

In 2004, Solidscape launched the BenchTop 3D printer series (T66BT and T612BT), a BenchTop-ready solution. The BenchTop series was based on DOS and did not include an external PC. The control software could run on the printer processing unit and the front-end software ModelWorks could be installed on the customer PC. Together with the BenchTop 3D printers, Solidscape has launched InduraCast and InduraFill model-making materials.

In 2006, Solidscape launched higher-performance BenchTop printers (T66BT2 and T612BT2).

In 2007, Solidscape launched the Windows platform-based BenchMark series of printers (T76, R66) including touchscreen functionality.

In 2009, Solidscape launched the preXacto series of printers (D76+, D66+) dedicated to dental applications, incorporating the proprietary SCP technology and DentaCast material.

In 2010, Solidscape launched the BenchMark (T76+, R66+), incorporating the proprietary SCP technology.

4.3   MCOR TECHNOLOGIES' SELECTIVE DEPOSITION LAMINATION (SDL)

Selective deposition lamination (SDL) or 3D paper printing was discovered by Dr. Conor and Fintan MacCormack in 2003. Dr. MacCormack first revealed the concept of 3D printing in 1986, when he was a high school student in Ireland watching a BBC special television show. Technology captured his imagination in a similar manner to how cars, rockets, computers, and space travel had already done. Originally, he saw the technology in person as he received his doctorate at Trinity College, Dublin. Unfortunately, the school's 3D printer was nothing more than a tease: Because of the very high price of the material, only one or two people could possibly print a component at the end of the year, beating the whole purpose of having the technology. When he began working with Airbus as an engineer, he had ample access to a 3D printer—access to which most students and engineers had been refused. It wasn't perfect. While 3D printer prices were decreasing, the cost of their materials was increasing. So Dr. MacCormack and his older brother, Fintan, a trained aircraft mechanic and electrical engineer, set out to develop a 3D printing machine with an operating cost so small that the technology would be available to everybody.

It was also essential to make the printer reliably sufficient for serious use in commercial settings, yet easy to use and without the toxic chemicals that so many 3D printers rely on. This vision has now become a reality in the co-founded MacCormacks Corporation, Mcor Technologies, which produces monochrome and full-color 3D printers that cost a fraction of any other 3D printing technology. The main explanation for this? While most technologies produce models of costly plastic or chemically infused powder, Mcor 3D printers use ordinary, inexpensive, and ubiquitous office paper as a building material.

SDL must not be confused with the old laminated object creation (LOM) technology. LOM used laser, laminated paper, and glue, so that everything, including the support material around the model, was glued together.

The model excavation was an awful experience, often resulting in a 3D part breakage. Mcor uses the blade to cut, and the 3D printer selectively deposits the adhesive only where it is required. This white paper will specify how a paper-based 3D printer produces a physical 3D model using the SDL process and document the specific features of an Mcor 3D printer that delivers MacCormack's vision. Generating digital file 3D printing starts with a 3D data file; Mcor 3D printers support an industry standard 3D product design file format, STL, as well as OBJ and VRML (3D color printing). All standard 3D computer-aided design (CAD) software products, including free programs like SketchUp, generate STL files. The completed designs offered for download are usually presented in the STL, as are files produced by scanning a physical object. Mcor 3D printers contain control software called SliceIT (Figure 4.2). SliceIT reads digital data and cuts the computer model into printable layers of paper thickness equivalent. The program also enables you to place the component, or multiple parts, inside the 3D printer's built-in chamber. SliceIT works on any standard 64-bit Windows (2000, XP, Vista, or Windows 7) PC with a dedicated Ethernet card (10/100 speed or higher) directly connected to a 3D printer.

IRIS also comes with an additional piece of software, called ColorIT, which is used in conjunction with SliceIT to apply color to digital 3D files (Figure 4.3).

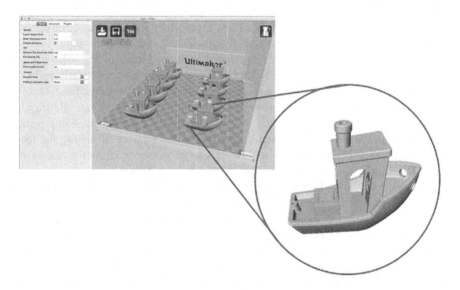

FIGURE 4.2 Mcor's SliceIT software reads digital data and cuts the computer model into printable layers of paper thickness equivalent. Compatible file formats include STL, OBJ, and VRML.

FIGURE 4.3   ColorIT applies color to the 3D digital files prior to slicing in SliceIT.

ColorIT can open a number of file formats: STL, WRL, OBJ, 3DS, FBX, DAE, and PLY. Once the file is inside ColorIT, its integrity can be checked to ensure that it is waterproof, but the primary purpose of ColorIT is to apply colors to digital files prior to SliceIT. Once the color has been applied, the model is exported as a WRL file, which is then imported into SliceIT for building planning and preparation.

The first sheet is manually mounted to the building wall. The location of the first sheet is not significant, as the first few pages are attached as a base layer before the actual portion is cut (Figure 4.4). Once the depth of the blade and the level of adhesion are correct, the doors are closed and the machine is ready to accept SliceIT data. The user selects print from the PC and SliceIT, and the 3D printer starts making the part.

The first thing that occurs is that a coating of adhesive is added to the top of the first manually placed board. The adhesive is selectively applied—thus the name SDL—"Selective." This means that a much higher adhesive density is deposited in the region that will become part of the adhesive and a much lower adhesive density is applied in the surrounding area that will serve as support (Figure 4.5). A new sheet of paper is fed from the paper feed mechanism into the printer and placed precisely on top of the freshly applied adhesive. The building plate is moved up to the heat plate and the pressure is applied. The pressure ensures a positive bond between the two sheets of paper (Figure 4.6). When the construction plate returns to the height of the construction, the adjustable tungsten carbide blade cuts one sheet of paper at a time, tracing the outline of the object to create the edges of the part (Figure 4.7). When this cutting sequence is done, the machine starts depositing the next layer of adhesive and the whole process is repeated until all the sheets of paper are glued together and cut and the pattern is finished. After the last layer has been completed, the portion can be removed from the construct chamber (Figure 4.8).

FIGURE 4.4   First sheet of paper is added to the platform.

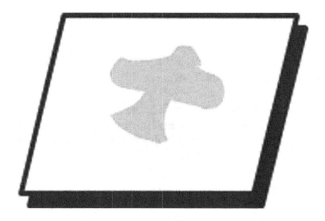

FIGURE 4.5   The adhesive is added to the select areas of the paper.

4.4   CUBIC TECHNOLOGIES' LAMINATED OBJECT MANUFACTURING (LOM)

There are a lot of different 3D printing processes available today, but have you ever heard of laminated object manufacturing? Laminated object manufacturing (or LOM) is really a very fast and affordable way to print 3D objects in a variety of materials. Material sheets are bonded together and cut in the right geometry according to the 3D model. Laminated object manufacturing is primarily used for rapid prototyping and not for production purposes.

Laminated object manufacturing is a 3D printing method developed by Helisys Inc. (now Cubic Technologies), but what happens in this process? Layers of material, plastic, or layers of paper are combined or laminated using heat and pressure. In LOM technology, the coated material is rolled onto the building platform. Usually,

FIGURE 4.6 Heat and pressure is applied to help bond the paper.

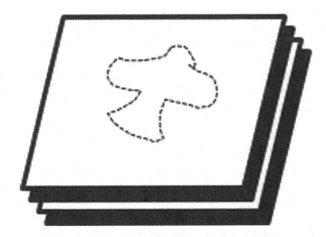

FIGURE 4.7 A tungsten carbide blade cuts the paper one sheet at a time along the cut line.

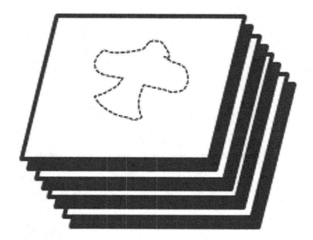

FIGURE 4.8 This process continues until the model is finished.

the material is coated with an adhesive layer and the feed roller heats to melt the adhesive. The layer is then bound to the previous one. The blade or laser is used to draw the geometry of the object to construct and draw crosses on the rest of the surface to facilitate the extraction of the final objects.

At the end of the building platform, there is a block composed of the final objects and the parallelepipeds to be removed. Objects printed using paper that have wood-like properties can benefit from sand casting finishes, while paper objects are usually sealed with paint or lacquer to keep moisture away. The technology has been introduced to the public by Cubic Technologies (formerly Helisys Inc.), which is proposing a plastic LOM system. Mcor has recently launched a paper-based system that adds color to the technology.

This technology is very versatile, as almost any material can be glued together. During this additive manufacturing process, layers of adhesive-coated paper, plastic, or metal laminates are successively bonded together. The most common material used is paper, because it is quickly cut. Plastic can also be used with a blade or laser during the cutting stage. The metal sheets are more unusual because the cutting stage is more complicated.

4.4.1 Working Process

In laminated object manufacturing, the paper is unwound from the feed roll (A) onto the stack and bonded to the previous layer by means of a heated roller (B). The roller melts a plastic coating on the bottom of the paper to form a bond. The profiles are tracked by an optical device that is placed on the X–Y stage (C). The method produces substantial smoke and a localized flame. Either a chimney or a carbon filtration system (E) is required and the construction chamber must be sealed. After cutting the geometric features of the layer, the excess paper is cut to separate the layer from the web. The extra paper of the network is wound on a roll (D) as shown in Figure 4.9.

4.4.2 APPLICATIONS

LOM machines are primarily used for rapid prototyping of plastic parts. Its low price and speed make it convenient to create prototypes, even though the items produced are far from the end-use pieces. Mcor proposes a particular type of LOM that they have named selective deposition lamination (SDL). It's a paper-based technology that adds color to the text. Paper sheets are printed in color, selectively bonded, and cut with a blade. The adhesive is only applied to the surface corresponding to the object, and the final object can be excavated more easily. In introduction, the addition of color allows this technology to compete with binder jetting technologies to produce multicolor artifacts, even if the consistency is not the same.

4.5 ULTRASONIC CONSOLIDATION

Ultrasonic consolidation or ultrasonic additive manufacturing was invented and patented by Dawn White. In 1999, White formed Solidica Inc. to sell UAM industrial

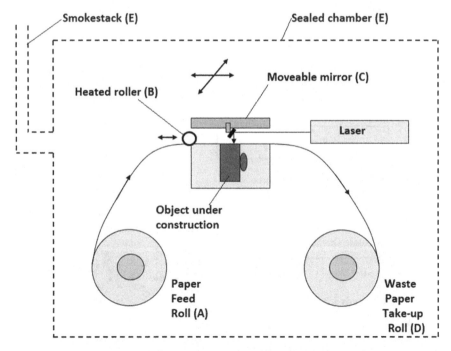

Laminated Object Manufacturing

FIGURE 4.9 Laminated object manufacturing process.

equipment—the Forming Machinery Suite. Around 2007, the Edison Welding Institute (EWI) and Solidica began working together to redesign the welding tooling to overcome bond consistency limitations and extend the process' weldable metals—so-called very high-power UAM. In 2011, Fabrisonic LLC was formed to commercialize the improved UAM process—SonicLayer machine suite. The process works by scrubbing metal foils together with ultrasonic vibrations under continuous pressure, i.e. classification of sheet lamination in additive manufacturing. Melting is not a mechanism of formation. Metals are instead connected to the solid state by disrupting surface oxide films between metals, i.e. ultrasonic metal welding mechanisms. CNC contour milling is used interchangeably with the additive stage of the process to incorporate internal features and add details to the metal part. UAM has the ability to combine multiple types of metals, i.e. dissimilar metal joints, with no or minimal inter-metallic formation and allows the embedding of temperature-sensitive materials at relatively low temperatures—usually less than 50 percent of the melting temperature of the metal matrix. As with most other additive manufacturing processes, UC creates objects directly from the CAD model. The file is then "sliced" into layers that result in the production of an STL file that can be used by the UC machine to build the required object layer by layer (Figure 4.10).

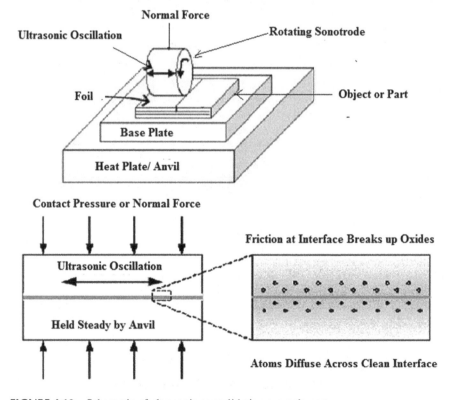

FIGURE 4.10 Schematic of ultrasonic consolidation example part.

The general manufacturing process is as follows:

- The base plate is placed on the anvil machine and set in place.
- The metal foil is then drawn under the sonotrode, which exerts pressure through normal force and ultrasonic oscillations, and bonded to the plate.
- This process is then repeated until the necessary area is covered by ultrasonic consolidated material.
- The CNC mill is then used to trim the excess foil from the part and to achieve the required geometry.
- Deposit and trim cycle shall be repeated until the specified height (usually 3–6 mm) is reached. A smaller finishing mill is used at this height to create the required tolerance and surface finish of the part.
- The deposit, trim, and finishing cycle shall continue until the finished object has been produced; at which point the anvil shall be removed and the finished article shall be removed from the base plate.

The advantage of this technology is the combination of different types of materials including aluminum and fiber optics. The disadvantage is that the process is limited to malleable metals that can be ultrasonically welded. The decision tree for this process is shown in Figure 4.11. Ultrasonic consolidation is a developing process for hybrid manufacturing and provides good capability for the combination of malleable materials and embedded electronics and fiber optics.

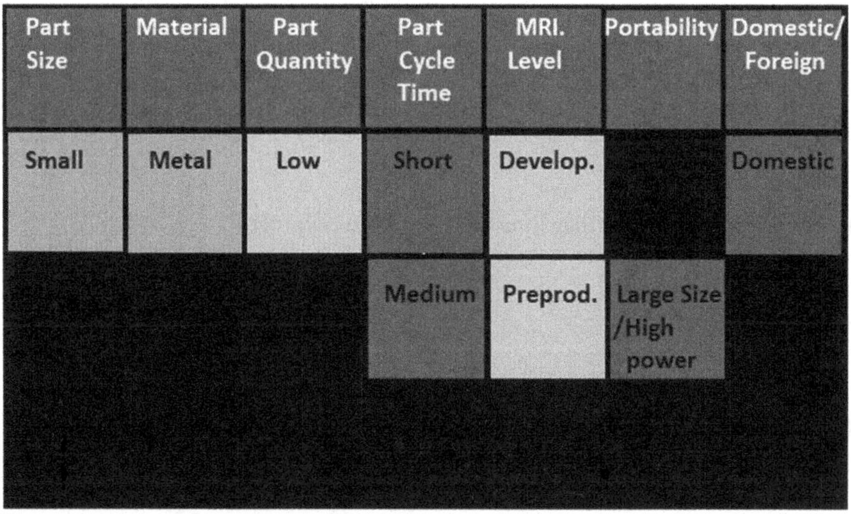

Part Size	Material	Part Quantity	Part Cycle Time	MRI. Level	Portability	Domestic/ Foreign
Small	Metal	Low	Short	Develop.		Domestic
			Medium	Preprod.	Large Size /High power	

FIGURE 4.11 Decision tree for ultrasonic consolidation.

4.6 EXERCISES

1. Explain Stratasys' fused deposition modeling (FDM).
2. Describe Solidscape's BenchTop system.
3. Discuss Mcor Technologies' selective deposition lamination (SDL).
4. Explain Cubic Technologies' laminated object manufacturing (LOM).
5. Briefly describe ultrasonic consolidation.

4.7 MULTIPLE-CHOICE QUESTIONS

1. Fused deposition modeling is a layer AM process that uses a
    _____ filament by fused deposition.
    a) Thermoplastic
    b) Thermoelastic
    c) Polylactic
    d) None of the above
    Ans: (a)

2. Solidscape was founded in _____.
    a) 1963
    b) 1986
    c) 1993
    d) 1999
    Ans: (c)

3. Selective deposition lamination (SDL) or 3D paper printing was discovered
    by Dr. Conor MacCormack and Fintan MacCormack in _____.
    a) 2003
    b) 2006
    c) 2007
    d) 2009
    Ans: (a)

4. LOM machines are primarily used for rapid prototyping of _____.
    a) Metallic parts
    b) Ceramic parts
    c) Wooden parts
    d) Plastic parts
    Ans: (d)

5. The main drawback to the use of FDM technology is _____.
    a) Quality
    b) Surface finish
    c) Speed of construction
    d) All of the above
    Ans: (c)

6. Which one is NOT related to the definition of rapid prototyping?
    a) Layer by layer
    b) Physical model
    c) From 3D CAD data
    d) Production line
    Ans: (d)

7. Which one of the following processes does NOT use laser?
    a) LOM
    b) SLA
    c) SLS
    d) FDM
    Ans: (d)

8. How many processes are there in the design process?
    a) 3
    b) 4
    c) 5
    d) 6
    Ans: (c)

9. Which of the following are processes in the RP cycle?
    a) Post-processing
    b) Transfer to machine
    c) Pre-processing
    d) All of the above
    Ans: (d)

10. Which of the following processes is available in color?
    a) SLA
    b) FDM
    c) MJM
    d) 3D printer
    Ans: (d)

5 Powder-Based Additive Manufacturing Systems

5.1 3D SYSTEMS' SELECTIVE LASER SINTERING (SLS)

Selective laser sintering (SLS) was invented and patented by Dr. Carl Deckard and academic adviser Dr. Joe Beaman at the University of Texas, Austin, in the mid-1980s, sponsored by DARPA. Deckard and Beaman participated in the resulting start-up company DTM, which was set up to design and manufacture SLS machines. In 2001, DTM was acquired by 3D Systems, the biggest competitor for DTM and SLS technology. The latest patent on Deckard's SLS technology was issued on 28 January 1997 and expired on 28 January 2014.

5.1.1 TECHNOLOGY

Selective laser sintering (SLS) printers from 3D Systems can perform rapid prototyping and produce high-resolution nylon parts up to seven times quicker than competing SLS 3D printers. The ability of SLS to manufacture many parts at once also makes the process a good choice for products needing strength and heat resistance from direct digital manufacturing (DDM) products. Additive production layer technology SLS involves the use of a high-power laser (e.g. carbon dioxide laser) to fuse small particles of plastic, metal, ceramic, or glass powder into a mass that has the desired three-dimensional shape. The laser selectively fuses the powdered material to the surface of the powder bed by scanning cross-sections created from a 3D digital representation of the object (e.g. from a CAD file or scan data). Once each cross-section is checked, one layer thickness lowers the powder bed, a new layer of material is applied to the rim, and the process continues until the portion is finished. The SLS system usually uses a pulsed laser because the final component density depends on peak laser power rather than laser duration.

The SLS system heats the bulk powder content well below its melting point in the powder bed, making it much simpler for the laser to raise the temperature of the selected regions to the melting point the rest of the way. Figure 5.1 shows the process of selective laser sintering (SLS).

In comparison to some other additive manufacturing techniques, such as stereolithography (SLA) and fused deposition modeling (FDM), which most often require specific support structures for the manufacture of overhanging designs, SLS does not need a different feeder for supporting material because the part being constructed is surrounded by uninterrupted powder at all times, which allows for the construction of a previous feeder.

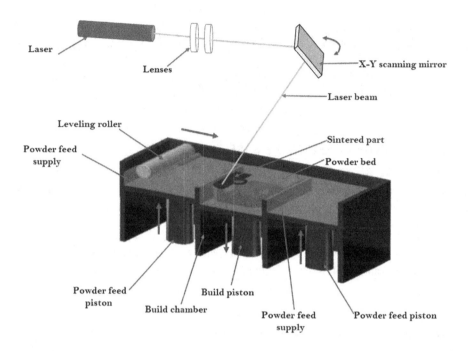

FIGURE 5.1   Selective laser sintering (SLS).

Also, since the chamber of the machine is always full of powder material, the manufacture of various parts has a much lower impact on the overall difficulty and price of the design because, by means of a method known as "Nesting," various parts can be positioned to fit within the limits of the machine. One design feature that should be noted, however, is that it is "impossible" for SLS to produce a hollow but fully enclosed element. This is because the uninterrupted powder cannot be drained inside the element.

5.1.2   Materials

Commercially available materials used in SLS come in powder form and include, but are not restricted to, polymers such as polyamides (PA), polystyrenes (PS), thermoplastic elastomers (TPE), and polyaryletherketone (PAEK). Due to its ideal sintering behavior as a semi-crystalline thermoplastic, polyamide is the most frequently used SLS material which results in parts with desirable mechanical properties. Polycarbonate (PC) is a material of great interest to SLS due to its high strength, thermal stability, and flame resistance; however, these amorphous SLS-processed polymers continue to result in components with reduced mechanical properties, dimensional accuracy, and are therefore restricted to applications where these are of low significance. Since the advancement of selective laser melting metal materials were not commonly used in SLS.

5.1.3   POWDER PRODUCTION

Powder particles are typically formed by cryogenic grinding in a ball mill at temperatures below the material's glass transition temperature which can be achieved by grinding with added cryogenic materials such as dry ice (dry grinding) or liquid nitrogen and organic solvent combinations (wet grinding). The process can result in spherical or irregular shaped particles that are as small as five microns in diameter. Powder particle size distributions are usually Gaussian and range from 15 to 100 microns in diameter, but they can be modified to match different layer thicknesses in the SLS process. Chemical binder coatings can be applied to post-process powder surfaces; these coatings aid in the sintering process and are particularly useful for the formation of composite material parts such as aluminum particles coated with thermoset epoxy resin.

5.1.4   SINTERING MECHANISMS

The sintering mechanism in SLS occurs primarily in a liquid state when the powder particles form a micro-melt layer on the surface resulting in reduced viscosity and a concave radial bridge between the particles forming, due to the material's reaction to lower surface energy, known as necking. In the case of coated powders, the laser's purpose is to melt the surface coating that acts as a binder. Solid state sintering is also a contributing factor but has a decreased effect, and occurs at temperatures below the material's melting point. The key driving force behind the cycle is again the material's reaction to lower its free energy level, which results in the particle-wide diffusion of molecules.

5.1.5   ADVANTAGE

One of the main advantages of SLS is that it does not require the support structures used by many other AM technologies to protect the design from collapsing during production. Since the product is in a powder bed, no support is required. This attribute alone, while also conserving materials, ensures that SLS is capable of creating geometries that no other technology can produce. In addition, we don't need to worry about damaging the part when removing the supports, so we can create complex interior components, so complete sections. As a consequence, we can save time on the assembly process. As with other AM techniques, there is no need to compensate for the issue of tool clearance—and therefore the need for joints—that subtractive methods frequently encounter. So we can make geometries that were previously difficult, minimize assembly time, and reduce poor joints. SLS is capable of producing highly durable components for real-world testing and mold construction, whereas other methods of additive manufacturing may become brittle over time. Since SLS parts are so durable, they compete with those made in conventional production methods such as injection molding, and are already being used in a range of end-use applications such as automotive and aerospace applications. Another major benefit of additive manufacturing with SLS is the ability to store and reproduce parts and die as 3D CAD data that will never corrode, lose transport, or require expensive storage. The designs are always available and ready to be produced when we need them, even if the original is not available.

5.1.6 APPLICATION

Due to its ability to easily manufacture complex geometries with little to no additional fabrication effort, SLS technology is commonly used in many industries around the world. It's most common application is early in the design cycle in prototype parts, such as investment casting patterns, automotive hardware, and wind tunnel models. In addition, limited production of SLS is increasingly being used to produce end-use parts for aerospace, military, medical, and electronic equipment. SLS can be used on the shop floor to rapidly assemble machines, jigs, and fixtures. Since the process requires the use of a laser and other costly, bulky equipment, it is not suitable for personal or residential use; however, applications in the field of art have been established.

5.2   3D SYSTEMS' COLORJET PRINTING (CJP) TECHNOLOGY

Binder jetting technology was founded at the Massachusetts Institute of Technology in 1993. One of the licensors of the technology was Zcorp or Zcorporation, which named built 3D printers depending on binder jetting and renamed Zprinting technology (as in Z-axis printing). When Zcorp was owned by 3D Systems, the market leader in stereolithography, in the early 2000s, the company began offering binder systems and renamed its ColorJet printing (CJP) technology because, with the addition of an inkjet head, it was possible to color the outer layer and thus the surface of the parts. ColorJet 3D printing, or CJP, is a 3D system 3D printing process that uses a core and a binder to create 3D objects.

This 3D printing process is famous for the printing of detailed, multi-color 3D parts for art, medical, consumer goods, and architecture and experimentation models. Parts can also be used as templates or master patterns for metal castings. The final pieces are delicate. Parts can be easily printed for a few days of turnaround.

5.2.1 TECHNOLOGY

As with many other rapid prototyping methods, the component to be printed with CJP/ZPrinting is made up of many thin parts of the 3D model. In ZPrinters, the inkjet-like printing head moves across the powder bed, selectively depositing the liquid binding material in the shape of the section. A fresh layer of powder is spread over the top of the model and the process is repeated. Once the pattern is finished, unbound powder is removed automatically. Parts can be produced on a ZPrinter at a rate of around 1 vertical inch per hour, making it one of the fastest technologies available today. In ColorJet printing, the outer edge of each layer is colored, resulting in complete colored final items. Figure 5.2 demonstrates the basic concept of ColorJet printing (CJP) technology.

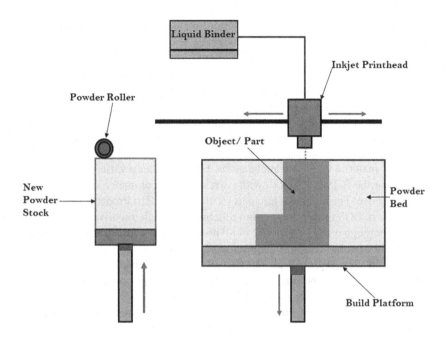

FIGURE 5.2 ColorJet printing (CJP) technology.

5.3   EOS'S EOSINT SYSTEM

5.3.1   ABOUT EOS (ELECTRO OPTICAL SYSTEMS)

Founded in 1989 and headquartered in Germany, EOS is a technology and market leader in additive manufacturing (AM) integrated e-manufacturing solutions, an industrial 3D printing process. EOS offers a modular portfolio of solutions, including systems, software, materials, and advancement materials as well as services (service, training, special application consulting, and support). As an industrial manufacturing technique, it allows for quick and scalable development of high-end components based on 3D CAD data at a quality level of the reproducible industry. It paves the way for a paradigm change to revolutionary technologies in product design and manufacture. This accelerates product development, gives versatility in design, optimizes component structures, and allows both lattice structures and functional integration.

5.3.2   EOSINT M 280 SYSTEM

The EOSINT M 280 is an upgraded and further improved version of the EOSINT M 270, the industry-leading tool for additive metal part manufacturing. On the basis of three-dimensional CAD data, it directly produces high-quality metal parts—fully automated—in just a few hours and without the need for instruments. The direct metal laser sintering (DMLS) method produces layer-by-layer parts by melting fine metal powder with a laser beam that allows extremely complex geometries such as free-form surfaces, deep grooves, and three-dimensional cooling channels to be created. Alternatively, the device is fitted with a 200 or 400 watt laser fiber. This type of laser offers extremely high beam efficiency and power stability, which can be controlled during the construction process using the laser power monitoring (LPM) method. The system operates in both argon and nitrogen protection atmospheres.

This allows the machine to handle a wide variety of materials: From light metals to stainless steel equipment and super alloys. Process software has been developed and improved over a number of years and includes a range of intelligent exposure techniques and features, which make it possible to refine and adapt the design process for a variety of material types and applications. EOS provides a variety of powder metal materials for the EOSINT M 280 with corresponding parameter sets that have been designed for use. They manufacture parts with uniform Part Property Profiles (PPPs).

In addition, EOS ensures maximum reliability through intensive process development and thorough quality assurance of all products. The capability of the system can be tailored to specific customer requirements with a range of choices and additional equipment. Integrated process chain management (IPCM) modules allow higher efficiency, higher quality, and improved user-friendliness, and can also be added at any time. The distinctive characteristics of the EOSINT M 280 system are the consistency of the components it manufactures and the ergonomically designed peripherals. Such features make the device the perfect manufacturing method for the economical batch-size, automated manufacturing of parts at all stages of the product life cycle. The machine is therefore ideally suited to the industrial environment (Figure 5.3).

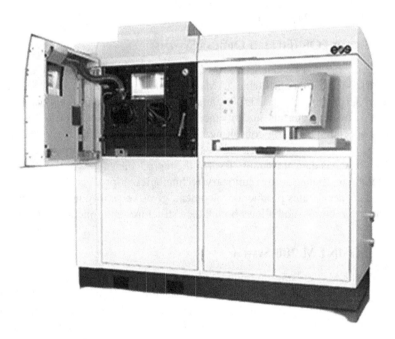

FIGURE 5.3   EOSINT M 280 system.

5.4 OPTOMEC'S LASER ENGINEERED NET SHAPING (LENS) AND AEROSOL JET SYSTEM

5.4.1 ABOUT OPTOMEC

Optomec is a privately owned, fast-growing provider of additive manufacturing systems. Optomec's proprietary aerosol jet systems for printed electronics and LENS 3D metal part printers are used by industry to reduce product costs and boost performance. These revolutionary printing techniques work together with the broad variety of used materials, from electronic inks to structural metals and even biological materials. Optomec has over 200 branded customers around the world, targeting manufacturing applications in the electronics, energy, life sciences, and aerospace industries.

5.4.2 LASER ENGINEERED NET SHAPING (LENS)

LENS systems are used for the repair and rapid manufacturing of metal parts in state-of-the-art materials such as titanium and stainless steel. The LENS MR-7 system can be used to quickly create materials of extremely high quality. The LENS MR-7 system offers a working envelope of 300 mm cubed, making it ideal for the manufacture or maintenance of smaller parts. LENS systems use high-power fiber laser energy to create structures directly from metal powders, alloys, ceramics, or composites, one layer at a time. Both powder feeders allow for the processing of gradient materials—each layer may have a different chemistry. This allows for the development and study of new materials at incredible pace. LENS systems are used for applications ranging from rapid alloy growth and practical prototyping to rapid manufacturing or repair over the entire product lifecycle (Figure 5.4).

5.4.3 HOW THE LENS SYSTEM WORKS

LENS systems use a high-power laser together with powdered metals to create entirely dense structures directly from a three-dimensional solid CAD model. The CAD model is automatically divided into a path of the tool which instructs the LENS machine to build the part. Under the supervision of software the component is designed layer by layer, which monitors a variety of parameters to ensure geometric and mechanical integrity. The LENS process is housed in an argon-purged chamber so that the level of oxygen remains below 10 parts per million to ensure no impurity is present during deposition. The metal powder is fed to the process by Optomec's proprietary powder feeding system which is capable of very accurately flowing small amounts of powder. When completed, the part is removed and can be heat-treated, hot-isostatic pressed, machined, or finished in any other way.

5.5 ARCAM'S ELECTRON BEAM MELTING (EBM)

Arcam supplies equipment for the manufacture of fully dense metal parts by electron beam melting (EBM). EBM technology uses a strong electron beam (4 kW power)

FIGURE 5.4   LENS MR-7 system.

to create part layer by layer of melting metal powder. The EBM process is conducted in a vacuum at an elevated temperature of 1000°C, resulting in stress-relaxed parts that have stronger material, mechanical, and chemical properties than cast and forged. The process is based on high-level energy utilization that delivers high melting capacity and high productivity.

The EBM process is planned primarily for the manufacture of refractory and resistant materials (tantalum, niobium, molybdenum, tungsten, vanadium, hafnium, zirconium, titanium) and their alloys. The EBM process is characterized primarily by high-speed production, rapid tooling, and complex geometry components with similar mechanical features to heat-treated materials. The Italian company Bticino has used EBM technology to produce light switch injection molding equipment in ABS plastic with a production volume of 1 million parts per year, with cobalt chromium alloy conformal cooling channels with high abrasion resistance and chemical corrosion. They were able to improve productivity, reduce cycle time, and lower the cost of production. Additionally, the efficiency of the manufactured parts has also been enhanced by improving the cooling system. The EBM tools in CoCr, a material with excellent wear and corrosion resistance properties, enabled BTicino to manufacture tools with extended geometrical freedom and longevity. Electron beam melting (EBM) is a new alternative for rapid manufacturing and prototyping of metal components. This technique is quickly gaining interest because of its ability to produce completely dense components, with properties comparable to those of wrought materials, at a cost and speed substantially lower than those of metal-based additive manufacturing methods.

EBM not only produces unparalleled strength-to-weight ratios, decreases the cost of raw materials, and decreases the weight of the parts, but also opens the door to new design configurations. EBM technology stands out for its ability to produce parts of titanium in hours versus days. For industries such as aerospace, this technology creates new opportunities for prototyping and low-volume component production. Time, costs, and challenges of machining or investment casting are eliminated, making titanium parts readily available for functional testing or installation on mechanical systems. EBM is patented by Arcam and distributed by Stratasys in the United States.

As the name suggests, EBM uses an electron beam to melt titanium powder. Additive manufacturing processes build parts on a layer-by-layer basis. After one layer of titanium powder is melted and solidified, the process is repeated for subsequent layers. Within the electron beam gun, the incandescent filament of tungsten and the electron cloud boil (Figure 5.5a). These electrons flow through the gun at about half the speed of light. Two magnetic fields are organizing and directing fast-moving electrons. The first act as a magnetic lens which focuses the beam on the desired diameter and the second magnetic field deflects the beam to the target point of the powder bed. When high-speed electrons strike metal powder, the kinetic energy is instantly transformed into thermal energy. Raising the temperature above the melting point, the electron beam quickly liquefies the titanium powder. Arcam A^2 (Figure 5.5b) developed by Arcam is capable of manufacturing parts up to 7.87 in. × 7.87 a.m. × 13.0 a.m. (200 mm × 200 mm × 330 mm in length). Parts produced with EBM are near-net shape like those made with metal-casting processes. As the electron beam completely melts the titanium, the liquefied metal conforms to the surrounding metal powder, which creates a surface finish similar to a precision sand casting; as a result, some light secondary grinding or grinding may be needed.

5.6 CONCEPT LASER'S LASERCUSING

The company, founded in 2000 by Frank Herzog, is one of the world's largest providers of machine and plant technology for 3D printing of metal parts. Since December 2016, Concept Laser has been part of GE Additive, the world's largest digital industrial group, General Electric. GE Additive was founded in 2016 and purchased, among others, 75% of the company's shares in Concept Laser.

5.6.1 CONCEPT LASER'S PATENTED LASERCUSING 3D-PRINTING TECHNOLOGY

The term LaserCUSING, also known as direct metal laser melting (DMLM), describes the technology: The fusion method develops components layer by layer

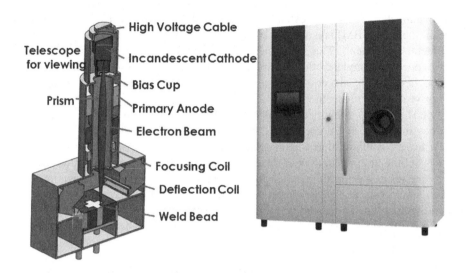

High Voltage Cable

Telescope for viewing

Incandescent Cathode

Bias Cup

Prism

Primary Anode

Electron Beam

Focusing Coil

Deflection Coil

Weld Bead

FIGURE 5.5   Electron beam melting (EBM).

using 3D CAD data. Concept Laser's patented Laser CUSING 3D-printing tech-nology is a metal additive manufacturing technique used to manufacture mechani-cally powerful and thermally robust components. The name of the technology, "LaserCUSING," is coined using the term laser "C" and the word "FUSING," which accurately defines the technology as a fusion method for the production of layer-by-layer metal components using a three-dimensional CAD model.

5.6.2   Working of LaserCUSING 3D-Printing Technology

LaserCUSING is a metal 3D-printing technology. This works similarly to powder bed fusion technology where the material is powdered and the laser is used to fuse the particles together. In the course of LaserCUSING, one layer of fine metal pow-der is dispersed around the building floor. The laser machine shines a high-energy fiber laser beam that is guided to the powder using a mirror redirector (scanner). The laser melts the powdered material. The adjacent particles merge together and solidify to develop a solid bond after cooling. Once the entire geometry of the layer is traced by the laser, the build platform shifts down and a second layer of powder spreads over the top of the first layer. This cycle continues until the entire portion is written. In addition to the patent for the process, Concept Laser also has a patent for what it calls stochastic control. This is what makes the whole system unique. The stochastic power of the slice segments (also abbreviated to as "islands") is processed successively. This proprietary method ensures a substantial reduction in stress in the manufacture of very large components (Figure 5.6, Figure 5.7).

The LaserCUSING layer construction process enables the development of close-contour cooling mold inserts and direct components for the jewelry, medical, dental, automotive, and aerospace industries. This refers to both prototypes and batch parts.

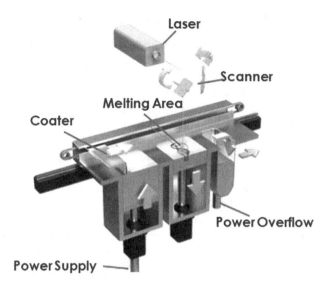

FIGURE 5.6 Concept Lasers' DMLM 3D printer.

FIGURE 5.7 Sample part 3D printed with LaserCUSING technology for polyshape.

5.6.3 Additional Features of LaserCUSING 3D-Printing Technology

- The complete printing process is carried out in an inert atmosphere so that the building chamber is not heated. This also minimizes the need to heat the material before the laser melting process. Process gasses are common when printing with different materials. Some of them are as follows:
- For tool steels and mold-making applications—nitrogen gas is used
- Titanium and aluminum—argon gas is used
- Selected steels—argon gas is used
- The same laser may be used for etching or engraving patterns/text on the metal component being printed.
- Often the vital pieces are more extruded and the laser surface finishes the top layer for a smooth finish.
- Nearly 99.5 percent content density is achieved.
- The layer resolution ranges from as low as 15 microns to as high as 500 microns.

5.6.4 Advantages of LaserCUSING 3D-Printing Technology

The LaserCUSING process offers many advantages, such as those listed below:

Green technology: The LaserCUSING process is a production method that produces almost no waste. Metal powder which has not been melted can be fully reused without the loss of any material for further processing. In fact, the laser process is almost emission-free. Due to the high degree of performance of the laser systems used at Concept Laser, the energy that is applied is efficiently transformed into working power.

Freedom of geometry: Complex geometry of components or geometry of components that cannot be generated by traditional means without the use of laser melting devices. There are no restrictions on the production of hollow or grid parts on the inside with this technology.

Near net shape: The design of components with close-net or ready-to-install geometry reduces manufacturing time and saves costs.

5.7 SLM SOLUTIONS' SELECTIVE LASER MELTING (SLM)

SLM Solutions from Germany launched the SLM500 HL machine in 2012, which uses double beam technology to increase the build rate up to 35 cm^3/hour and has a build volume of $500 \times 350 \times 300$ mm^3. Two sets of lasers are used in this machine, each set having two lasers (400 W and 1000 W). This means four lasers scan the powder layer simultaneously.

Selective laser melting (SLM) is a powder bed AM technology in which parts are fabricated layer by layer using the action of a high-energy beam on a powder bed. In this process, the powders are fully melted and solidified. The process is very similar to the SLS process but the energy of the beam is much higher and the process is performed under a controlled atmosphere. SLM is currently very popular for fabrication of metallic parts (Figure 5.8).

Several attempts have been made to produce ceramic materials directly without sintering or post-processing using SLM technology. The Fraunhofer Institute of Laser Technology was able to produce completely ceramic net-shaped specimens of almost 100 percent density without post-processing by the complete melting of the powder with the Nd:YAG laser beam. Nevertheless, ultra-high preheating was used to prevent crack formation during the construction process. That makes the process very difficult. A eutectic mixture of $Al2O3 \pm ZrO2$ was used to lower the melting point of the material. While a good prototype has been shown, the mechanical properties of the parts are not comparable to ceramics developed by traditional methods. Figure 5.9 shows the images of produced parts using this method.

SLM has been identified for sintering refractory ceramics such as $ZrB2$ for high-temperature applications. Zirconium metal was used as a binder and $ZrB2$-Zr cermets with a density of more than 95% and appropriate mechanical properties were obtained.

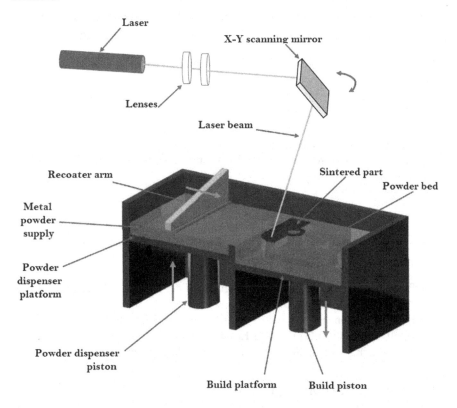

FIGURE 5.8 Schematic of selective laser melting (SLM).

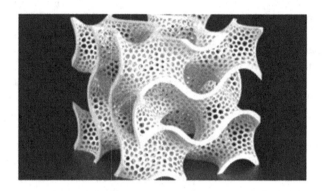

FIGURE 5.9   Ceramic parts manufactured by the selective laser sintering process.

In general, the following are the key drawbacks of SLM technology for the manufacture of ceramic materials:

- Balling effect due to low conductivity of ceramic material
- Extensive cracking due to high temperature fluctuations during processing
- Limited flowability of ceramic after melting and inability to form a dense layer
- Porosity formation that is severely detrimental to the mechanical properties of the component
- High preheating of the chamber is needed to reduce temperature gradient and subsequent cracking

However, the direct additive manufacture of ceramic is still a major advantage of the process. Extensive research is needed to minimize the harmful aspects of the process defect.

5.8   EXERCISES

1. Explain 3D Systems' selective laser sintering (SLS).
2. Discuss 3D Systems' ColorJet printing (CJP) technology.
3. Describe EOS's EOSINT systems.
4. Briefly describe Optomec's laser engineered net shaping (LENS) and aerosol jet system.
5. Explain the concept of Arcam's electron beam melting (EBM).
6. Briefly discuss Concept Laser's LaserCUSING.
7. Briefly write about SLM Solutions' selective laser melting (SLM).

5.9   MULTIPLE-CHOICE QUESTIONS

1. What is the full name of SLS?
   a) Selective laser simulator
   b) Sintering laser simulator

    c)   Selective laser sintering
    d)   Stereolithography laser sintering
    Ans: (c)

2.   What is the other name for multi-jet modeling?
    a)   FDM
    b)   PolyJet
    c)   3D printer
    d)   Extrusion
    Ans: (b)

3.   Which one of the following is the design process?
    a)   Build
    b)   Concept
    c)   Pre-processing
    d)   Transfer to machine
    Ans: (b)

4.   What is the file format for prototyping?
    a)   .prt
    b)   .slt
    c)   .stl
    d)   .iges
    Ans: (c)

5.   Which CAD software cannot be used to create data for prototyping?
    a)   CREO
    b)   CATIA
    c)   NX UniGraphics
    d)   Adobe Illustrator
    Ans: (d)

6.   Which one of the following processes is subtractive prototyping?
    a)   5-axis CNC milling
    b)   Fused deposition modeling
    c)   Multi-jet modeling
    d)   Stereolithography apparatus
    Ans: (a)

7.   In which of the following processes are the input materials in solid form?
    a)   SLA
    b)   SLS
    c)   FDM
    d)   MJM
    Ans: (c)

8.   In which of the following processes are the input materials in liquid form?
    a)   LOM
    b)   SLS

    c)   FDM
    d)   MJM
    Ans: (d)

9.   In which of the following processes are the input materials in powder form?
    a)   LOM
    b)   SLS
    c)   FDM
    d)   MJM
    Ans: (b)

10. Which material is NOT available for the LOM process?
    a)   Paper
    b)   Plastic
    c)   Metal
    d)   Glass
    Ans: (d)

11. Which of the following processes uses the extrusion concept?
    a)   SLA
    b)   SLS
    c)   FDM
    d)   MJM
    Ans: (d)

12. Which model of 3D printer is available in PERDA-TECH?
    a)   Z310
    b)   Z450
    c)   Z510
    d)   Z650
    Ans: (c)

13. Which of the following is NOT the color binder of a 3D printer?
    a)   Cyan
    b)   Black
    c)   Magenta
    d)   Yellow
    Ans: (b)

14. Which of the following is the process of the pre-processing stage?
    a)   Remove support
    b)   Checking 3D CAD data
    c)   De-powdering loose material
    d)   Dip in binder to strengthen the part
    Ans: (b)

15. What is the infiltrant used to strengthen parts in the Z510 machine?
    a) Water
    b) Paint
    c) Epson salt
    d) Color bond
    Ans: (d)

6 Materials in Additive Manufacturing

6.1 CHOOSING MATERIALS FOR MANUFACTURING

Advances in technology, along with subsequent material innovations, have had a significant effect on the way 3D printing is treated and relied on by engineers, designers, and manufacturers during development and production. Through additive manufacturing, technology transforms the material through heat, light, or other guided energy.

There are four main types of materials corresponding to specific technologies: Photopolymers, powdered thermoplastics, filament thermoplastics, and metals. Materials must be adapted to the application in order to produce successful results. The properties of any material become increasingly important as the product moves from conceptual and functional prototyping to production. However, the material properties can only be assessed when the manufacturing process is considered. It is the combination of the material and the process that defines the characteristics of the material. For example, an alloy manufactured by die casting has different properties when it is molded by injection of metal.

In the same way, thermoplastics will have different properties if they are injection molded or CNC machined. Additive manufacturing (AM) or 3D printing is unique. This varies from all other manufacturing processes because the material properties and characteristics of the parts that it manufactures are different, even though it uses almost the same metal or thermoplastic. In terms of material properties, it is not a matter of being better or worse; it is simply important to recognize that the effects will be different. Recognizing that there is a distinction, the following information will help identify and eventually select materials from three commonly used industrial 3D printing processes: Direct metal laser sintering (DMLS), selective laser sintering (SLS), and stereolithography (SLA). Picking the right material comes down to a number of factors. It may be difficult to keep track of each property required for the material of the component and to decide if the material is suitable for the manufacture of the goods.

Below are some material considerations which help to find the suitable material for the manufacturing of products.

6.1.1 APPLICATION

When choosing a material and 3D printing method for your project, make sure your material suits the certifications and/or key features needed for the application. Depending on where the part is in the life cycle of the product, you will have distinct durability needs. For example, a concept model may need to reflect the look and feel of the end product, but it does not necessarily need to have the same durability of the end product. Once you have identified the needs of your usage, the Material Wizard allows you to filter through all of our available materials that meet these requirements by clicking on the Key Characteristics tabs or adjusting the material properties' sliders to specific measurements.

6.1.2 Asthetics

As stated above, 3D printing materials are often inseparable from technology, and each technology delivers parts with a variety of resolutions. PolyJet builds parts with the smallest layer and the full color of the CMYKW, resulting in highly detailed cosmetic parts. It provides expert finishing services; some materials convert better than others into sanding, polishing, and painting. You can use the Key Characteristics buttons in the Material Wizard, such as High Resolution/High Detail, Clear/Translucent, and Flexible, to filter through the best materials for highly esthetic parts.

6.1.3 Function

3D printing materials are subjected to stringent testing in order to react to the type of stress that they can withstand and to the degree of the demanding environment in which the material can excel. Filter through Key Characteristics such as Toughness, Flame Retardance, Impact Resistance, and Product Properties to find the product that suits the primary feature of your application.

6.1.4 Certifications

Certain 3D printing materials provide biocompatibility, sterilization capabilities, FDA skin contact certifications, flame smoke and toxicity certifications, chemical resistance, or other certifications that may be essential to your project. It's important to ensure that your material can produce what you need when selecting a material and 3D printing process for your project. Stratasys Direct Manufacturing, a printing service provider with ISO 9001 and AS 9100 certifications, will ensure strict content and technical specifications are met.

While our collection of available 3D printing materials is expansive, it isn't all-inclusive. The breadth of 3D printing material options is growing, bridging gaps between prototype and end-use production. Machine manufacturers and third-party materials developers have seen huge implications of future evolution to accommodate new and innovative applications.

Listed below are the materials used for industrial 3D printing.

6.1.4.1 Nylon

Nylon (known as polyamide) is a thermoplastic synthetic linear polyamide and is the most popular material in plastics. Because of its versatility, longevity, low friction, and corrosion resistance, it is a well-known 3D printing filament. Nylon is also a common material used in making clothes and accessories. Nylon is suitable for use when constructing delicate and complex geometries. It is used primarily as a filament on 3D printers in FDM (fused deposition modeling) or FFF (fused filament fabrication). This material is inexpensive and considered one of the hardest plastic materials.

6.1.4.1.1 Distinct Characteristics
- Nylon is known for its strength and durability.
- It has an excellent ratio of strength to flexibility.
- Nylon has a very small warpage.

This type of material can be smoothly dyed or colored.

6.1.4.1.2 Disadvantages
- Because nylon is hydroscopic, it should be kept dry.
- It has a shelf life of 12 months.
- Such material can shrink during cooling, so printing may be less accurate.
- The suitability of printers also varies.

6.1.4.2 ABS (Acrylonitrile Butadiene Styrene)
ABS is a thermoplastic that is commonly used as a 3D printer filament. It is also a material generally used in personal or household 3D printing and is a go-to material for most 3D printers.

6.1.4.2.1 Distinct Characteristics
- It is one of the most accessible and cheap materials for 3D printing.
- ABS is highly available and has a wide variety of colors.
- This material has a longer lifespan compared to nylon.
- It is also mechanically strong.
- This material is not suitable for hobbyists. It is mainly used by producers and engineers looking for high-quality prototype production.

6.1.4.2.2 Disadvantages
- When going to print, it requires a heated bed.
- Because ABS materials have a high melting point, they tend to experience warping when cooled when printing.
- The type of filament is a non-biodegradable toxic material that releases toxic fumes with a terrible smell at high temperatures.

6.1.4.3 Resin
Resin is one of the most widely used 3D printing materials. It is generally used in technologies such as SLA, DLP, multi-jet, or CLIP. There are different types of resins that can be used for 3D printing, such as castable resins, hard resins, flexible resins, etc.

6.1.4.3.1 Distinct Characteristics
- It can be used in a number of applications.
- It has a low shrinkage rate.
- Resin materials are of high chemical resistance.
- This is a solid and delicate material.

6.1.4.3.2 Disadvantages
- It's expensive.
- This type of filament will also expire.
- Due to its high photo-reactivity, it needs to be stored safely.
- It may cause premature polymerization when exposed to heat.

6.1.4.4 PLA (Polylactic Acid)

PLA or polylactic acid is derived from renewable sources such as sugar cane or cornstarch. It's also called "black plastic." It is often used in primary and secondary schools because it is safe to use and easy to print. This is also used for FDM screen printing.

6.1.4.4.1 Distinct Characteristics
- PLA is easy to print as it has a low warping effect.
- It can be printed on a cold surface, too.
- It can be printed with sharper edges and features compared to ABS material.
- This material is available in a variety of colors.

6.1.4.4.2 Disadvantages
- PLA materials are not very durable and can deform when exposed to intense heat.
- This type of material is less durable.

6.1.4.5 Gold and Silver

Nowadays, 3D printing can be done using gold and silver. These filaments are robust materials and are processed in powder form. These materials are generally used in the jewelry sector. These metals use either the DMLS (direct metal laser sintering) or the SLM method for printing.

6.1.4.5.1 Distinct Characteristics
- It has a high electrical conductivity.
- It has proof of heat.

6.1.4.5.2 Disadvantages
- Gold and silver printing is expensive.
- It consumes a lot of effort and energy to get it right.
- Both gold and silver are difficult to operate with lasers due to their high reflectivity and high thermal conductivity.

- Since these materials require extremely high temperature printing, a regular FDM 3D printer is not suitable for use.

6.1.4.6 Stainless Steel

Stainless steel is printed by fusion or laser sintering. There are mainly two possible technologies that could be used for this material. It could be DMLS or SLM technologies. Since stainless steel is all about strength and detail, it is perfect for use in miniatures, bolts, and key chains.

6.1.4.6.1 Distinct Characteristics

- Stainless steel can be heat treated to improve strength and hardness.
- It works well in high strength applications.
- It provides a strong resistance to corrosion.
- Ithashigh ductility.

6.1.4.6.2 Disadvantages

- Building time for 3D printing using these metals is much longer.
- Stainless steel printing is expensive.
- The size of printing is limited.

6.1.4.7 Titanium

Titanium is the strongest and lightest 3D printing material. It is used for a process called direct metal laser sintering. This metal is mainly used in high-tech areas such as space exploration, aeronautics, and the medical industry.

6.1.4.7.1 Distinct Characteristics

- This offers greater flexibility and design resolution.
- It gives production accuracy to industrial designers.
- It has an average roughness of the surface.
- Titanium is also biocompatible and resistant to corrosion.

6.1.4.7.2 Disadvantages

- 3D titanium printing is expensive.

6.1.4.8 Ceramics

Ceramics is one of the latest technologies used in 3D printing. Glass is more durable than metal and plastic as it can withstand intense heat and pressure without cracking or warping. Therefore, this type of material is not susceptible to corrosion like other metals or wears away like plastics. This material is commonly used in binder jetting, SLA (stereolithography), and DLP (digital light processing) technologies.

6.1.4.8.1 Distinct Characteristics

- It has high-precision components with a smooth, glossy surface.
- It also has resistance to acid, heat, and lees.
- It has a wide range of colors.

6.1.4.8.2 Disadvantages
- Ceramics require a high amount of temperature to melt.
- This is not appropriate for glazing and cooking processes.
- Since it is fragile, it has limitations on printing objects with enclosed and interlocking parts.
- It is not suitable for the assembly of pieces.

6.1.4.9 PET/PETG
Like nylon, PET or polyethylene terephthalate is also one of the most widely used plastics. This material can be used in thermoforming process. It can also be combined with other materials such as glass fiber to produce engineering resins. PETG is used for 3D printing. This is a modified version of PET where the G stands for "glycol-modified." As a result, a filament that is less fragile, clearer, and easier to use than PET is formed. This filament is applicable to FDM or FFF technologies.

6.1.4.9.1 Distinct Characteristics
- The material is robust.
- It is impact-resistant and recyclable.
- It can also be sterilized.
- It has excellent adhesion to the layer.
- It has the combined functionality of ABS (temperature-resistant, stronger) and PLA (easy to print).

6.1.4.9.2 Disadvantages
- The material may be weakened by UV light.
- It's prone to scratching.
- More testing of 3D printing parameters is needed.

6.1.4.10 HIPS (High Impact Polystyrene)
HIPS or high impact polystyrene is plastic filaments which are used for support structures in FDM printers. It is equivalent to ABS when it comes to ease of use. The only dissimilarity is its ability to dissolve. HIPS is highly soluble to a liquid hydrocarbon called limonene.

6.1.4.10.1 Distinct Characteristics
- It has excellent mechanical properties. It can also be used to create complicated structures.
- It's very smooth and lightweight.
- It is completely waterproof and impact resistant.
- It's very cheap.

6.1.4.10.2 Disadvantages
- Produces strong fumes. Thus, it is recommended to be used in a ventilated area.
- Without continuous heat flow, this material will clog up the nozzle and distribution tubes of the printer.

6.1.4.11 Thermoplastics

Thermoplastics are suitable for functional applications, including the manufacture of end-use components and functional prototypes. They have good mechanical properties, high impact, abrasion, and chemical resistance. They may also be filled with carbon, glass, or other additives to improve their physical properties. 3D print engineering thermoplastics (such as nylon, PEI, and ASA) are widely used in the production of end-use parts for industrial applications.

SLS components have better mechanical and physical properties and higher dimensional precision, but FDM is more affordable and has shorter lead times.

	Typical 3D printing thermoplastics
SLS	Nylon (PA), TPU
FDM	PLA, ABS, PETG, Nylon, PEI (ULTEM), ASA, TPU

The pyramid below shows the most common 3D printing thermoplastic materials. As a rule of thumb, the higher the material in the pyramid, the better its mechanical properties and the more difficult it is generally to print (higher cost) (Figure 6.1).

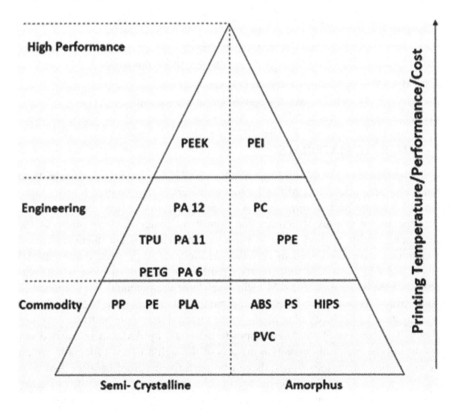

FIGURE 6.1 Common thermoplastic materials for 3D printing.

6.1.4.12 Thermosets (Resins)

Thermosets (resins) are widely used for applications where esthetics are important, as they can produce parts with smooth injection-like surfaces and fine details. Normally, they have high stiffness but are more porous than thermoplastics, so that they are not ideal for practical applications. Specialty resins are used for engineering applications (mimicking the properties of ABS and PP) or dental extensions and implants. Material jetting produces components with superior dimensional accuracy and generally smoother surfaces, but at a higher cost than SLA/DLP. Both processes use similar acrylic-based photocurable resins.

Typical 3D printing thermosets (resins)	
Material jetting	Standard resin, digital ABS, durable resin (PP-like), transparent resin, dental resin
SLA/DLP	Standard resin, tough resin (ABS-like), durable resin (PP-like), clear resin, dental resin

6.1.4.13 Metals

Metal 3D-printed components have outstanding mechanical properties and can be worked at high temperatures. The freeform 3D printing capabilities make it suitable for lightweight applications in the aerospace and medical industries. DMLS/SLM components have excellent mechanical properties and tolerances, but binder jetting can be up to ten times cheaper and can generate much larger parts.

Typical 3D printing metals	
DMLS/SLM	Stainless steel, titanium, aluminum
Binder jetting	Stainless steel (bronze-filled or sintered)

6.2 MULTIPLE MATERIALS

The unique aspect of the additive manufacturing (AM) technology is the ability to produce multi-material parts. Multiple types of materials may be used for the manufacture of a single part. Components with specially tailored functionally graded, heterogeneous, or porous structures and composite materials were some of the achievements of this method. A broad variety of materials, such as metals, plastics, and ceramics, have been used in various AM methods to produce multi-material goods in order to satisfy the existing requirements of the industry that would otherwise not have been met. All AM techniques have the potential to be applied to multiple materials in nature. In addition, a number of studies have been conducted to investigate the possibility of applying multi-material development to various AM methods. The process of making composite materials by AM can be carried out either during the material deposition process or through a fusion process in which the combination of various materials can be performed before or after AM as a previous or successive stage of the production of a component. Composite processes can be used to produce heterogeneous scaffolds and functionally graded materials

(FGMs). One of the features of AM processes is the development of customized gradient multi-phase or porous materials. As a result, different properties can be achieved within a single integrated component. In addition, when making a component with composite materials, the required properties of the included materials can be combined while compensating for some of their limitations.

6.2.1 MULTI-MATERIAL AND COMPOSITE ADDITIVE MANUFACTURING METHODS

Various modifications of the AM and the combination thereof with other manufacturing methods for the production of multi-material or composite products are mentioned below. The subjects are mostly categorized as process adaptations for the implementation of multi-materials, different materials used in these processes, and AM hybrids and other manufacturing methods.

6.2.1.1 Stereolithography Methods

Stereolithography (SLA) has been developed on the basis of photopolymerization phenomena and often includes the implementation of a light source for bonding photocurable resins mixed with other materials for the manufacture of solid composite parts. Multi-material SLA processes have been carried out by successive application and washing of various types of photopolymerizable resins in single or multiple vat configurations (Figure 6.2). The versatility of these approaches has spread from the manufacture of electronic parts to biomedical implants in a variety of fields. Typically, the liquid precursor infiltration method has been used for the production of ceramics and their composites. A porous component is immersed in a liquid infiltration material in this method. As a result of the infiltration of the precursor into the pores of the ceramic component, a huge variety of microstructures such as gradient, partially or fully dense materials, and also an increase in density and mechanical properties of the compact powder can be achieved. In the SLA of ceramic materials, a ceramic suspension including photocurable liquid resin is used to produce the green part of the de-ceramic component. Drying and other debinding processes are then carried out in order to achieve a component with high density and minimal defects which has been infiltrated by an SLA process of aluminum ceramics by immersing the printed products in liquids of various altering components to create a multi-phase composite by filling the interconnected porosities with infiltrated materials. A photo-initiator material was added to the aluminum suspension to make it UV-curable and suitable for the SLA process. Debinding and subsequent infiltration were followed by precipitation (Figure 6.2b). As a result of the infiltration process, the hardness of the part increased but the fracture strength decreased.

6.2.1.2 Binder Jetting Methods

Binder jet printing deposits binder materials on powder bed for selective bonding of powder materials layer by layer for the construction of three-dimensional parts. In binder jetting, additional extractable powder materials can be used in the binding process to achieve the desired percentage of materials in different layers of parts. Subsequently, additional materials can be extracted to achieve the desired porous or

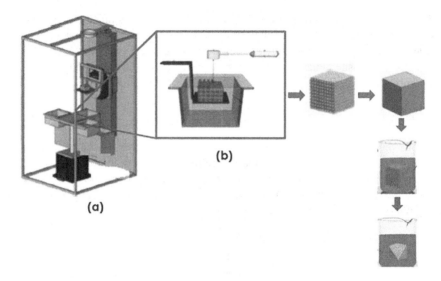

FIGURE 6.2 (a) Multiple vat setup for stereolithography of multi-material parts; (b) SLA followed by infiltration and precipitation process.

functionally graded material (FGM) by means of extraction techniques such as solvent materials. The application of infiltration processes to ceramic materials developed by AM was used to build multi-phase composites. As for the binder jet process, the process uses the nature of the binder jet's porous products. While the binder material is cured to form a solid part, it burns off and leaves the porosity in the final component. The main purpose of the infiltration procedure is to fill the porosity with other functional materials in order to obtain a fully dense part. Ceramic metallic composites can be produced by bonding, curing, and sintering ceramic powder materials under special conditions in order to achieve a solid homogeneous structure and by immersing them in a molten metal bath to fill the porosities with metal. The sintering temperature of the printed ceramic component can affect the density and volume fraction of the metal phase and, as a result, the microstructure and the mechanical properties of the final part. The submerging time of the sintered ceramic component in the molten metal is another conceivable parameter that can change the properties of the products as the more sintered ceramic parts are deposited in the molten metal; apart from the phase of penetration, the more metallic material permeates the ceramic particles and the properties of the products will have a higher resemblance to the pure metallic parts made of the molten metal. This technique has also been used for the production of metallic composites. For one test, it was shown that the thickness of the printed component has a critical influence on the different properties of the final component as it alters the microstructure and chemical composition of the product.

6.2.1.3 Extrusion-Based Printing Methods

Extrusion-based AM uses an extrusion nozzle to position materials on a substrate to print the desired part layer by layer. Low-temperature deposition (LDM) is an extrusion-based AM method capable of manufacturing multi-material components with custom-built porosities (Figure 6.3). Because the LDM process operates at low temperatures, bioactivity and biocompatibility of biomaterials can be maintained. This method is capable of creating scaffolds with functionally graded or composite materials. Biomaterial scaffolds consisting of synthetic and natural polymers in tissue engineering applications such as bone and cartilage structure scaffolds or nerve and vascular tissues are the main focus of this manufacturing technique. The technique typically involves depositing a mixture of materials on an ultra-low temperature platform with a sterilized syringe followed by a freeze drying process to remove the solvent material and the micro pores formed by phase separation. In order to improve their mechanical and physiochemical properties, LDM made scaffolds with a core-shell composite structure. The inner and outer feedstock tube and the nozzle head were assembled together to simultaneously extrude the core and the sheath material. A multi-nozzle LDM system using disposable syringes has been used to manufacture scaffolds with gradient biomaterials and tissue engineering functions.

Fused deposition modeling (FDM) uses filaments comprising thermoplastic polymers which are melted and extruded on the desired substrate layer by layer through the nozzle. One of the polymers manufactured by this method is polyvinylidene fluoride (PVDF), which has recently been extended to energy harvesting systems. In ceramic AM FDM, a polymer and ceramic composite can be used as feedstock material to enhance the physical and mechanical properties of the final product introduced by FDM to produce PP and TCP composite scaffolds with optimized interconnected porosities. Various scaffolding architectures and compatibility of

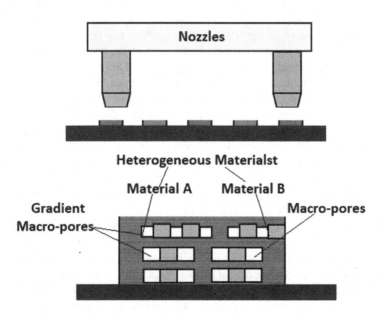

FIGURE 6.3   Schematic of multiple nozzle extrusion based AM of gradient materials.

scaffolding materials for vitro cell culture were evaluated and the results suggested the process' ability to generate biomaterial components. The quality and durability of the composite filament for the extrusion process is a crucial factor in achieving the desired properties of the products. Other AM-related parameters, such as raster gap and width as well as slice thickness, are effective in the structure of the scaffold and therefore in the mechanical properties and dispersion of the porosity. The use of sacrificial polymeric mixture or UV-curable resins mixed with ceramic particles is common in many AM methods for the production of ceramic components. The additive mixture plays the role of binder in the AM process of producing a green product and is removed from the part by a subsequent debinding process. However, each of these approaches usually suffers from certain barriers of its own, such as excessive material consumption, low sintering density, the need for additional steps, lack of functionality, and limitations in the dimensions of the components produced. The addition of photopolymerizable dispersion to the raw material of the AM-based extrusion process was attempted to take advantage of the robustness of the green product with UV-cured resin and the economical AM-based syringe process. UV-light irradiation is carried out during the printing process and UV-resin is removed by a traditional sintering operation. However, deficiencies such as partial polymerization of the layers and cracking of the component due to a high shrinkage ratio and during sintering have yet to be dealt with.

6.2.1.4 Material Jetting Printing Methods

The principle of inkjet printing (IJP) usually involves the deposition of a jet stream of droplets or particles of materials so that the materials can combine and attach together. Various actuating systems for the ejection of droplets have been studied, including thermal bubble form, piezoelectric nozzle head, pneumatic diaphragm actuator, and the use of an electrical field to produce ink flow. The use of multiple nozzles or the mixing of raw materials for printing on the printhead enables the production of multi-material, multi-phase composites and FG materials. A combination of tape casting technique and IJP was applied to the manufacture of electrolytes and micro-battery electrodes. Flexible super capacitors have been manufactured by IJP with graphene-based materials on metal film. Ink-jetted dielectrics were also used for printing insulators for packaging of electronic embedding devices and for securing crossovers of metallic materials. IJP was used to store photoactive layer materials for solar cells and multiple photodetector materials, both in combination with applications such as spin coating deposition. For the manufacture of flexible phototransistors, a combination of reverse offset printing, IJP, and bar coating was used. IJP is responsible for the polymeric active semiconductor and conductive polymer for gate electrodes in this method. Multi-jet modeling (MJM), also known as material jetting system or PolyJet printing of materials, uses multiple jet nozzles to store photopolymers for part structures that are immediately cured after deposition, and gel-like wax materials for sacrificial support structures. A schematic representation of the method is shown in Figure 6.4. This AM technique is capable of manufacturing components with greater resolution and geometric complexity, such as microfluidic devices with narrow gaps and high aspect ratios.

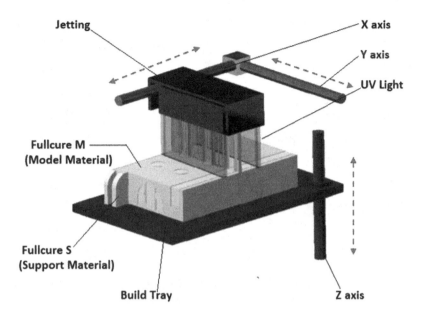

FIGURE 6.4 Schematic of PolyJet 3D printing.

The implementation of the electrospinning technique in syringe-based AM production was carried out in order to achieve micro-/nano-scale fiber diameters. Using electro-spinning in biomaterial AM manufacturing methods will lead to improved mechanical and biological properties and encourage cell proliferation in tissue engineering. A few of the limitations of electrospinning methods, such as buckling and jet stream coiling, have been resolved by performing a hybrid process with melt-electrospinning. A combination of AM-based syringe and wet-spinning technique, which includes the extrusion of a polymer mixture filament into a precipitating bath to solidify and manufacture continuous composite fibers, has been performed to produce microfibers for microporous composite scaffolds. Since high temperature, high voltage, or toxic solvents are not needed, wet spinning allows the development of fibers, including biomolecules, which makes it ideal for bioprinting. Figure 6.5 displays the configuration of electro-spinning and wet-spinning processes.

6.3 METAL AM PROCESSES AND MATERIALS

A variety of different technologies used in the metal additive manufacturing systems are available today. Systems may be categorized by the energy source or the manner in which the material is joined, e.g. by means of a binder, laser, heated nozzle, etc. It is also possible to classify the group of materials being processed, such as plastics, metals, or ceramics. The feedstock condition, the most common of which is solid (powder, wire, or sheet) or liquid, is often used to describe the process.

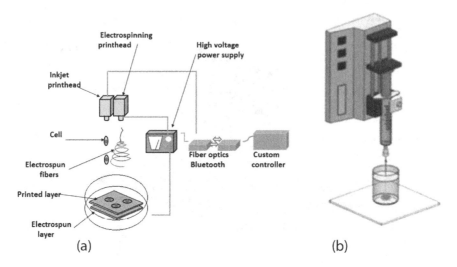

FIGURE 6.5 (a) Hybrid of inkjet printing and electrospinning; (b) Setup of wet-spinning method.

6.3.1 POWDER-BED SYSTEMS

Almost every powder-bed AM system uses a powder deposition method comprised of a coating mechanism to spread a powder layer onto a substrate plate and a powder reservoir. Usually the layers are 20 to 100 μm thick. Once the powder layer has been distributed, a 2D slice, known as 3D printing, is bound together or melted using an energy beam applied to the powder bed. For the second example, the energy source is usually one high-powered laser, but the state-of-the-art systems can use two or more lasers of distinct power in an inert gas atmosphere.

Direct process powder-bed systems are known as laser melting processes and are available commercially under various trade names such as selective laser melting (SLM), laser cusing, and direct metal laser sintering (DMLS). The only exception to this process principle is the electron beam melting (EBM) process, which uses a full-vacuum electron beam. The melting cycle is repeated by slice, layer by layer, until the last layer is melted and the pieces are finished. It is then removed from the powder bed and processed as required (Figure 6.6).

6.3.2 POWDER-FED SYSTEMS

Since the powder-fed systems are using the same feedstock, the manner in which the material is applied layer by layer varies considerably. The powder flows through the nozzle, which is melted from the beam right on the surface of the treated part (Figure 6.7)

Powder-fed systems are also known as laser cladding, direct energy deposition, and laser metal deposition. The process is highly efficient and is based on the automated deposition of a layer of material with a thickness of between 0.1 mm and a few

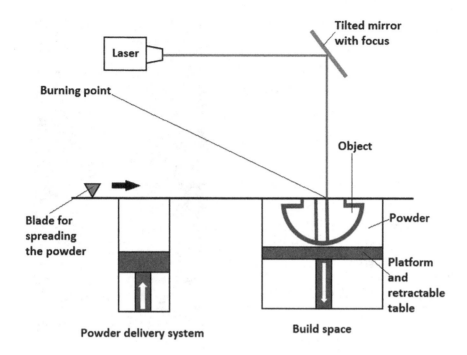

FIGURE 6.6 Schematic diagram of the selective laser melting (SLM) powder-bed process.

centimeters. Some features of this process are the metallurgical bonding of the cladding material with the base material and the lack of undercutting. The process is different from other welding techniques in that a low heat input penetrates the substrate.

One of the advancements in this technology is the laser engineered net shaping (LENS) powder delivery system used by Optomec. This technique allows the addition of material to an existing part, which means that it can be used to repair expensive metal components that may have been damaged, such as chipped turbine blades and injection molding tool inserts, offering a high degree of flexibility in the clamping of the parts and the "coating" materials.

The companies that offer systems based on the same principle are: BeAM from France, Trumpf from Germany, and Sciaky from the United States. An interesting approach to the hybrid system is the one offered by DMG Mori. The combination of the laser cladding theory and the 5-axis milling method opens up new fields of application in many industrial branches.

6.3.3 DIRECT METAL LASER SINTERING (DMLS) MATERIALS

DMLS uses pure metal powders to manufacture products with properties that are generally accepted to be equal to or better than those of wrought materials. Due to rapid melting and solidification in a small, continuously moving area, DMLS can produce differences in grain size and grain boundary that have an effect on

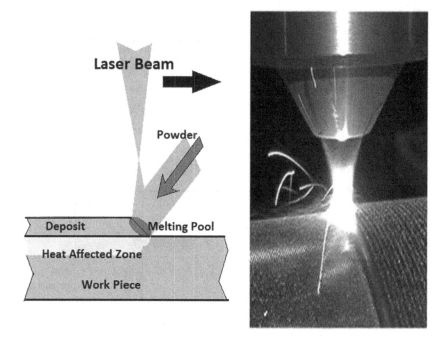

FIGURE 6.7   Schematic of the laser cladding process and the laser cladding process in action (right).

mechanical efficiency. Research is ongoing to characterize grain structures that can change with laser parameters, post-construction heat treatment, and hot isostatic pressing. The results, however, are not widely available. At the end of the day, this difference will become an advantage when the grain structure can be manipulated to offer different mechanical properties to the part.

(a) **Stainless steel** is a commonly used DMLS material and is available in two grades at Protolabs: 17-4 PH and 316L. Select 17-4 for its significantly higher tensile strength (190 ksi vs. 70 ksi), yield strength, and hardness (47 HRC vs. 26 HRC), but recognize that it has far less elongation at break (EB) than 316L (8 percent vs. 30 percent), which means it will be less malleable. Both 17-4 and 316L offer corrosion resistance, but 316L is better suited to acid resistance. 316L is also more temperature-resistant than 17-4. Note that 17-4 can be heat treated to change mechanical properties, whereas 316L is only usable in a stress-relieved state.

(b) **DMLS aluminum (Al)** is equivalent to the 3000 series of alloys used in casting and die casting processes. It's made of AlSi10Mg. Al has an outstanding strength-to-weight ratio, good temperature, and corrosion resistance, and good fatigue, creep, and rupture strength.

Compared to die-cast aluminum series 3000, the Al properties for tensile strength (36 ksi to 43 ksi) and yield strength (30 ksi to 32 ksi) much surpass the average values. However, the elongation at break is significantly lower (1 percent vs. 11 percent) relative to the 3000 series aluminum average.

(c) **DMLS titanium (Ti-64 ELI)** is most commonly used for aerospace and defense applications due to its strength-to-weight ratio, temperature resistance, and resistance to acid/corrosion. It is also used for medical applications. Versus Ti grade 23 annealed, the mechanical properties are almost identical with a tensile strength of 130 ksi, a break elongation of 10%, and a hardness of 36 HRC.

(d) **Cobalt chromium (CoCr)** is one of two superalloys of DMLS that tend to be used for aerospace and medical specialty applications. CoCr has an exceptional EB (20 percent) and is resistant to creep and corrosion. Versus ASTM F-75 CoCr (depending on heat treatment), DMLS CoCr provides moderate material properties (DMLS vs. F-75): Tensile strength of 130 ksi vs. 95–140 ksi, EB 20 percent vs. 8–20 percent, yield strength of 75 ksi vs. 65–81 ksi, and hardness of 25 HRC vs. 25–35 HRC. Of all DMLS metals, CoCr has the best biocompatibility—which requires additional biocompatibility treatment outside of Protolabs—making it ideal for medical applications such as dental implants.

(e) **Inconel 718 (IN718)** is a nickel chromium superalloy used in high-temperature applications such as aircraft engine components. DMLS IN718 parts have an impressive operating temperature range of -423°F to 1300°F combined with excellent corrosion resistance and good fatigue, creep, and rupture strength.

DMLS IN718 has a higher tensile strength (180 ksi vs. 160 ksi) and comparable yield strength (133 ksi vs. 160 ksi) than the conventional IN718. Its EB, however, is half that of conventionally processed IN718 (12 percent vs. 25 percent) (Table 6.1).

6.3.4 SELECTIVE LASER SINTERING (SLS) MATERIALS

SLS uses thermoplastic powders, primarily polyamide (PA), to manufacture usable parts with higher strength and impact strength than those provided by SLA, as well

TABLE 6.1
Metal 3D Printing Materials Comparison Chart

Material	Tensile Strength	0.2% Yield	Elongation	Hardness
Aluminum (ALSI10MG)	37.7 ksi	20 ksi	10%	47.2 HRB
Cobalt Chrome	130 ksi	75 ksi	20%	25 HRC
Inconel 718	180 ksi	150 ksi	6–12%	35.5 HRB
Titanium (TI-6AL-4V)	129 ksi	164 ksi	10%	40 HRC
Stainless Steel 17-4PH	190 ksi	170 ksi	8%	40–47 HRC
Stainless Steel 316L	70 ksi	25 ksi	30%	76.5 HRB–25.5 HRC

as high HDTs (351°F to 370°F). The tradeoffs are that SLS lacks the smoothness of the surface finish and the fine specifics of the features available with SLA.

(a) **DuraForm HST Composite** is a fiber-filled PA that is equivalent to a 25 percent mineral-filled PA 12. The HST fiber content significantly improves strength, stiffness, and HDT. Compared to other SLS and SLA options (excluding ceramic-filled materials), HST has the maximum tensile strength, flexural modulus, and impact strength and maintains an elevated HDT. This makes HST a great choice for practical applications where temperatures above 300°F can be present. However, the material is somewhat fragile, with an EB of 4.5 percent. Also, consider that there is a significant delta in the Z-axis values, such as injection molded fiber-filled materials.

(b) **PA 850** Black delivers ductility and flexibility with a tensile modulus of 214 kpsi and an EB of 51 percent, all without sacrificing tensile strength (6.9 ksi) and temperature resistance (370°F HDT). These characteristics make PA 850 a common general purpose material and the best solution for making living hinges for limited testing.

Compared to the PA 11 injection molded average, PA 850 has a higher HDT (370°F vs. 284°F) with equal tensile strength and stiffness. However, its EB, although the strongest in all AM plastics, is 60 percent lower than that for PA 11 molded plastics.

Another aspect that distinguishes PA 850 is its uniform, deep-black color. Black has a high contrast, which makes the features pop, and hides dirt, grease, and grime. Black is also desirable for optical applications because of low reflectivity.

(c) **ALM PA 650** is a balanced, economical, general-purpose material. PA 650 is stiffer than PA 850 (tensile modulus 247 ksi vs. 214 ksi) and has a higher tensile strength (7.0 ksi vs. 6.9 ksi). Although it's EB is half that of PA 850, it is still one of the top performers in terms of ductility at 24 percent. PA 650 is loosely comparable to the average PA 12 injection molding properties. It has similar stiffness but approximately half the tensile strength and the EB. Nevertheless, the HDT is slightly higher: 351°F vs. 280°F.

(d) **PA 615-GS** is a polyamide powder filled with glass spheres that make it rigid and dimensionally stable. However, the glass filler makes PA 615-GS brittle, significantly reducing impact and tensile strength. Glass spheres also make PA 615-GS parts much heavier than any other AM material.PA 615-GS mimics the average value of glass-filled injection molded nylon. Compared to 33 percent of glass-filled nylon, HDT is lower at 350°F vs. 490°F with much lower tensile strength (4.5 ksi) and EB (1.6 percent) (Table 6.2).

6.3.5 STEREOLITHOGRAPHY (SLA) MATERIALS

SLA uses photopolymers, ultraviolet (UV) cured thermoset resins. It offers the wide range of materials with a wide range of tensile strengths, tensile and flexural

TABLE 6.2
Nylon 3D Printing Materials Comparison Chart

Material	Tensile Strength	Tensile Modulus	Elongation	Heat Deflection
PA 650	6.96 ksi	247 ksi	24%	186–351°F
PA 850 Black	6.1–6.95 ksi	214 ksi	51%	180–370°F
DuraForm HST	4.5–7.35 ksi	164 ksi	2.7–4.5%	276–363°F
PA 615-GS	4.5 ksi	170 ksi	1.60%	273–350°F

modules, and EBs. Note that the impact strengths and HDTs are lower than those of normally utilized injection molded plastics. There are also choices for color and hardness in the range of materials. Combined with a good surface finish and high resolution, SLA can produce parts that resemble injection molding in terms of performance and appearance.

(a) **Accura Xtreme White 200** is a widely used SLA material. In terms of flexibility and strength, it falls between polypropylene and ABS, making it a good choice for snap fits, master patterns, and challenging applications. Xtreme is a robust SLA material with a very high impact strength (1.2 ft.-lb./in.) and a high EB (20 percent) while it is medium in strength and stiffness. However, its HDT (117°F) is the lowest of all SLA materials. Compared to the average value for injection molded ABS, Xtreme may have a slightly higher tensile strength (7.2 ksi vs. 6.0 ksi) but slightly lower EB (20 percent vs. 30 percent). Under a flexing load, the Xtreme is 26% less rigid and its impact strength is 70% lower.

(b) **Somos WaterShed XC 11122** provides a rare combination of low moisture absorption (0.35%) and near-colorless transparency. Secondary operations will be required to ensure that the material is completely clear, and will also retain a very light blue hue afterward. Although good for general-purpose applications and pattern-making, WaterShed is the best choice for flow-visualization models, light pipes, and lenses.

WaterShed's tensile strength and EB are among the best in 3D-printed, thermoplastic-like materials, making it tough and durable. WaterShed offers a slightly higher tensile strength (7.8 ksi vs. 6.0 ksi) compared to average injection molded ABS values, but falls short in EB (20 percent vs. 30 percent) and HDT at 130°F vs. 215°F.

(c) RenShape **7820** is another alternative for the prototyping of injection molded ABS parts. Not only does it imitate the mechanical properties of ABS, its deep black color and the glossy surfaces of the top profile offer the appearance of a molded part, while the layer lines can be visible in the side profile. RenShape 7820 also has a low absorption of moisture so that the pieces are more dimensionally stable.

Compared to other SLA materials, there are midrange values for all mechanical properties. Compared to the average ABS values (injection molded), the tensile strength is slightly higher (7.4 ksi vs. 6.0 ksi), but the EB is lower (18 percent vs. 30 percent) and the HDT is 124°F vs. 215°F. The most significant departure from ABS is a low impact force of 0.91 ft.-lb./in.

(d) Accura **60** is an alternative to both RenShape SL 7820 and WaterShed XC 11122 when stiffness is required. Like RenShape SL 7820, this material produces clear, crisp details; like WaterShed, this material offers translucency. However, this material sacrifices ductility with 29 to 36 percent lower EB and 10 to 44 percent lower impact strength. Furthermore, Accura 60 has a high absorption rate of moisture, which can impair dimensional stability. Accura 60 has a high tensile strength of 9.9 ksi and a high tensile strength of 450 ksi. Compared to the average values for injection molded, 10 percent glass-filled polycarbonate, it has equivalent tensile strength and flexural modulus, 25 percent higher EB but 80 percent lower impact strength.

(e) Somos **9120** is the best choice of SLA resins when polypropylene-like parts are needed. This material is the most flexible SLA choice, with a flexural modulus of 210 ksi and it is the most ductile, with an EB of 25%. It also has the second highest impact strength (1.0 ft.-lb./in.) for SL products. In direct comparison to the average values of injection molded polypropylene, 9120 has comparable tensile strength (4.7 ksi), tensile modulus (212 ksi), flexural modulus (210 ksi), and impact strength (1.0 ft.-lb./in.). The only deviation from the shaped PP is a 75% lower EB.

(f) **Accura SL 5530** provides a strong, rigid part with high temperature resistance. In addition, a post-cure thermal option can increase HDT from 131°F to 482°F. 5530 has the highest tensile and flexural moduli (545 ksi and 527 ksi) of all unfilled SLA materials and the second highest tensile strength (8.9 ksi). The post-cure can, however, make 5530 less durable, resulting in an impact strength of only 0.4 ft.-lb./in. and the EB is 2.9 percent. Without post-curing thermal heat, 5530 maintains its tensile strength and becomes more flexible. Already, the EB is rising by 50%.Compared to injection molded thermoplastics, 10% of the glass-filled polycarbonate is the closest match. For post-cure thermal, 5530 has comparable tensile strength and flexural modulus (compared to average values) with 66% higher HDT. However, the impact strength and EB are much lower for 5530 (81% and 72%, respectively).

(g) **MicroFine Green**™ is custom designed in Protolabs to deliver the highest degree of detail—0.002 inch features are possible—and the strictest tolerance available from any SLA content. The material is used to produce small parts, generally less than 1 in. by 1 in. by 1 in.

In terms of mechanical properties, MicroFine Green™ falls in the midrange of SLA materials for tensile strength and modulus (6.5 ksi and 305 ksi, respectively) and in the low end for impact strength and EB (0.46 ft.-lb./in. and 6 percent, respectively).

MicroFine Green™ has hardness (329 ksi vs. 333 ksi) and tensile strength (6.5 ksi vs. 6.0 ksi) comparable to injection molded ABS, but MicroFine™ has a lower HDT than ABS of 138°F vs. 215°F.

TABLE 6.3
Metal 3D Printing Materials Comparison Chart

Material	Tensile Strength	Elongation	Heat Deflection
Accura Xtreme White 200	6.3–7.2 ksi	7–20%	108–117°F
Somos WaterShed XC 11122	6.8–7.8 ksi	11–20%	115–130°F
RenShape 7820	5.2–7.4 ksi	8–18%	125°F
Accura 60	8.41–9.86 ksi	5–13%	118–131°F
Somos 9120	4.4–4.7 ksi	15–25%	126–142°F
Accura SL 5530	6.8–8.9 ksi	1.3–4.4%	131–482°F
MicroFine	6.5 ksi	6.10%	122–138°F
Metal Plated	14–29 ksi	1%	122–516°F

Protolabs offers another proprietary material, SLArmor, which combines nickel plating with Somos NanoTool parts to offer an alternative to die-cast aluminum. Plating increases the tensile strength of NanoTool to 14.5 ksi to 29 ksi, depending on the percentage of metal thickness. HDT is dramatically increased over NanoTool with a spectrum of 122°F to 516°F and, relative to die-cast aluminum, HDT is greater than 500°F with a tensile strength of 43.5 ksi (Table 6.3).

6.4 COMPOSITE MATERIALS

3D printing, also known as additive manufacturing, is a process used to make a solid object from a three-dimensional digital model, usually by laying down several successive thin layers of a material. In recent years, developers have found ways to print composite materials, such as carbon fiber and fiberglass, which improve the durability and structure of 3D-printed items. While the technology of 3D printing composite materials is not mainstreamed yet, many manufacturers have begun to use it to print everything from vehicle and aircraft parts to the walls of future buildings. Composite materials are preferred above some traditional materials because of their strong, lightweight, and more affordable properties. Composite material is a material consisting of two or more constituent materials with very different physical and chemical properties. When combined, they produce characteristics that are different from each variable.

The two constituent materials can be divided into two categories: The matrix (binder) and the reinforcement. A composite material requires a matrix and reinforcement. The matrix is used to support the reinforcement material by surrounding it so that it can maintain its relative position. The reinforcement relies on its special mechanical and physical properties to strengthen the properties of the matrix. This creates an enhanced synergy that would otherwise not exist in the individual constituent materials. It also enables product designers to optimize the combination of composites. In spite of the relatively high cost, fiber-reinforced composite materials have become prevalent in high-performance products for their lightweight yet powerful properties. They can withstand tough loading conditions in aerospace components, boat and scull hulls, bicycle frames, and racing car bodies. They are also used in fishing rods, storage tanks, pool panels, and baseball bats. The new Boeing 787 construction, including wings and fuselage, is primarily made of composites. These are also gaining prominence in the field of orthopedic surgery.

6.4.1 GENERIC COMPOSITE MATERIALS

Concrete is the most common artificial composite. Usually, it consists of loose stones (aggregate) and is attached to a cement matrix. The term "plywood" is derived from thin layers known as "plies" of wood veneer that are bonded together. Plywood sheets are bonded to each other with wood grains rotated up to 90 degrees to achieve reinforcing properties. Plywood is known as an engineered wood in the board family which includes medium-density fiberboard (MDF) and particleboard (chipboard). The two composite components of plywood are resin and wood fiber sheets, consisting of long, strong, and thin cellulose cells. Cross-graining, which alternates grains, reduces the tendency of the wood to divide by nailing at the edges and decreases expansion and shrinkage to improve dimensional stability. It also provides more consistent power across all panel directions. In order to reduce warping, an odd number of folds is generally present to help with the balance. Cross-graining an odd number of composites also increases the strength of the wood so that it becomes difficult to bend perpendicular to the grain direction of the surface ply.

(a) Fiberglass is a common type of fiber-reinforced plastic (FRP) reinforced with glass fiber. Glass fibers are embedded in the material by sporadically arranging, compressing into a flat sheet, or interweaving into a fabric. The plastic matrix is either a thermoset polymer matrix (epoxy, polyester, or vinyl ester) or a thermoplastic matrix. Fiberglass is a more cost-effective and flexible alternative than carbon fiber; it is also stronger than many metals and can be molded into complex shapes. Fiberglass is used in a wide range of applications such as aircraft, automobiles, bathtubs, and enclosures, boats, castings, cladding, exterior door leathers, hot tubs, pipes, roofing, septic tanks, surfboards, swimming pools, and water tanks. Fiberglass is sometimes referred to as glass-reinforced plastic (GRP), glass-fiber reinforced plastic (GFRP), or GFK (Glassfaserverstärkter Kunststoff). One important distinction to be made is between glass fiber and fiber-reinforced plastic, because sometimes the glass fibers themselves can be called "fiberglass."

(b) Carbon fiber is an incredibly strong and lightweight fiber-reinforced composite made up of carbon fiber. While it tends to be costly, it is generally applied to industries requiring high strength-to-weight ratios and rigidity, such as aerospace, automotive, civil engineering, and sports goods. Its implementation is increasingly advanced in the various consumer and technical applications. The carbon fiber binder is usually a thermoset resin such as epoxy, but can also be thermoplastic polymers such as polyester, vinyl ester, or nylon. Other fibers such as aramid (Kevlar, Twaron), titanium, ultra-high-molecular-weight polyethylene (UHMWPE), or glass fibers as well as carbon fiber can also be included in the composite. Carbon fiber is sometimes referred to as graphite-reinforced polymer or graphite-reinforced polymer. Many common names include carbon fiber reinforced plastic or carbon fiber reinforced thermoplastic (CFRP, CRP, and CFRTP) or generally referred to simply as carbon fiber, carbon composite, or even carbon.

6.4.2 COMPOSITE MATERIALS FOR 3D PRINTING PROCESSES

(a) The aluminum content is a combination of polyamide and a very low amount of gray aluminum powder. Laser sintering is a technique used to create complex, conceptual, functional, or series models that are strong and relatively rigid. Models made of aluminum can also absorb small impacts and withstand a certain amount of pressure under bending conditions. It is the ideal material for novice designers seeking affordability, maximum design freedom, and increased printing capabilities. In addition, it delivers a higher stiffness than polyamide would have on its own and features an aluminum appearance. The surface is mildly porous with a gritty, granular appearance and can be finished in a natural gray matt state or painted with a variety of colors.

(b) Fiber-reinforced nylon enables 3D printing of engineering parts that are as strong as aluminum at the cost of plastic. It is specifically designed with aluminum strength and has a higher strength-to-weight ratio than 6061-T6 aluminum, and is up to 27 times stiffer and 24 times stronger than ABS. Fiber-reinforced nylon enables users to refine their printed end-use pieces, in particular functional prototypes and tests, structural parts, jigs, fixtures, and devices, while concentrating on strength, stiffness, weight, and temperature resistance. It is not suitable for small parts with complex designs.

(c) Full-color sandstone provides photo-realistic full-color models and sculptures and is particularly suited to architectural models, life-like sculptures, gifts and memorabilia, and fine arts. Essentially, it is made of gypsum and has a colored texture on the surface. The brittleness of the material limits its use as functional parts or as intricately designed parts. This sandstone material is sometimes referred to as multicolor.

6.5 BIOMATERIALS, HIERARCHICAL MATERIALS, AND BIOMIMETICS

6.5.1 BIOMATERIALS

Biomaterials rapid prototyping (RP), recently recognized as additive manufacturing (AM), has emerged as a revolutionary technology, promising to turn science into medical therapy. RP is a layer-by-layer manufacturing process that directly translates computer data such as computer-aided design (CAD), computer tomography (CT), and magnetic resonance imaging (MRI) into three-dimensional (3D) objects. RP innovations play a significant role in the biomedical industry, such as anatomical models for training/planning, recovery, dentistry, personalized implants, drug delivery devices, tissue engineering, and organ printing. The incorporation of biomaterials and rapid prototyping techniques has been an exciting avenue in the development of biomaterial implants over the last decade. Biomaterials are used to develop functional regeneration of different tissues to improve human health and quality of life. Biomaterials may be natural or synthetic. Additive manufacturing (AM) is a new material processing approach for creating layer-by-layer parts or prototypes directly from a computer-aided design (CAD) file. The combination of additive manufacturing and biomaterials is very promising, particularly for patient-specific clinical applications. The challenges of AM technology, along with related material issues, need to be addressed in order to make this approach workable for broader clinical needs.

6.5.1.1   Biomedical

3D printing implemented in the medical sector has been available for a number of years through various applications. Organ transplantation has difficulties and 3D tissue engineering-based jet printing offers a possible solution. Some research defines organ printing as a fast prototyping computer-aided 3D-printing technology based on the use of layer-by-layer deposition of cell and/or cell aggregates into a 3D gel with sequential maturation of the printed structure into perfumed and vascularized living tissue or organ. Bio-printing is a desirable way to create tissues and organs in hospitals. Successful implantation depends on compatible materials. A variety of biomaterials can be found, such as curable synthetic polymers, synthetic gels, and naturally derived hydrogels. Prosthetics is the first biomedical field to use 3D printing and presents a number of achievements. We can quote a patient's skull anatomy reproduced by 3D printing for pre-surgical use in manual implant design and production, and by enhancing the stability of the fixation of the custom-made total hip prosthesis and restoring the original biomechanical characteristics of the joint. Several applications combine some compostable or allergenic scaffolding with cellular bio-printing to create personalized biologic prosthetics that have the potential to act as a transplantable replacement tissue. New articles have shown that the medical 3D printing market could reach $983.2 million by 2020.

6.5.2   Hierarchical Materials and Biomimetics

Biological materials with hierarchical architectures (e.g. macroscopic hollow structure and microscopic cellular structure) bring unique inspiration for the design and manufacture of advanced biomimetic materials with outstanding mechanical performance and low density. Most traditional biomimetic materials benefit only from bio-inspiring architecture on a single-length scale (e.g. microscopic material structure), which significantly limits the mechanical performance of the resulting materials. There is great potential to improve the mechanical performance of biomimetic materials by using a bio-inspiring hierarchical structure. The three-dimensional (3D) ink-based printing technique for the development of ultra-light biomimetic hierarchical graphene materials (BHGM) with exceptionally high rigidity and durability is demonstrated. By simultaneously engineering 3D-printed macroscopic hollow structures and constructing an ice-crystal-induced cellular microstructure, BHGMs can achieve high elasticity and stability at compressive strains of up to 95%. Multiscale finite element analyses indicate that the hierarchical structures of BHGMs effectively reduce the macroscopic strain and turn microscopic compressive deformation into rotation and bending of interconnected graphene flakes. This 3D printing study underscores the great potential that exists for the assembly of other functional materials into hierarchical cellular structures for various applications where high stiffness and low density resilience are simultaneously required.

Nature creates a wide range of materials to suit specific structural functions, such as protection and mechanical support. In spite of the mechanically inferior components, the outstanding mechanical properties of biological materials can be achieved through hierarchically structured structures. In this context, a hierarchical structure refers to a special arrangement of structural elements in such a way that the geometry and properties of the structure change from one length scale to another. Bone, mother-of-pearl, tooth enamel, antler, sea sponge exoskeletons, diatoms, alpha-helix-based protein filaments, geckos' feet, lotus leaf, and spider silk are just a few examples of materials using inferior components and structural hierarchy to achieve remarkable collections of properties and functionalities. Biopolymers and minerals are basic building blocks of nearly all biological materials. In terms of mechanical properties, the polymers are soft and deformable, whereas the latter on it becomes rigid and brittle. Soft polymers provide ductility and flexibility to biological materials, while rigid minerals provide stiffening and reinforcement. Nature therefore provides numerous examples of how soft and rigid ingredients can be combined to achieve higher levels of ductility and strength at the same time. It contrasts, however, with the approach that human beings have taken for a long time, i.e. by enhancing the mechanical properties of a single material by materials science techniques and/or by inventing new materials, such as steel, copper, and silver, which are not found in natural materials. However, the combination of materials with mechanical properties complementing each other is not sufficient to produce high-performance materials. For example, numerous attempts to imitate the structure and performance of mother-of-pearl, the shiny inner layer of many mollusk shells, have produced interesting materials, but with properties and mechanics that are still inferior to the natural model. The level of mechanical properties of the components achieved in natural mother-of-pearl is far beyond that of synthetic counterparts. Behind this performance is a highly sophisticated microstructure in which features are well defined and controlled over several length scales, which is the result of a bottom-up approach to manufacturing. In this approach, small building blocks are self-assembled (a process in which a disordered element compound is rearranged into a well-organized structure), mineralized (in the case of hard biological materials), and shaped larger elements. The cycle of hierarchical structuring has made this transition from lower to higher levels possible. "Growing" from small to large scale through this technique allows the material a high degree of flexibility to adapt/optimize the structure to different length scales according to specific functional requirements. By contrast, the traditional approach to the manufacture of engineering materials is top-down, where the material is removed to achieve the desired shapes and components. For example, in order to produce a punch, the steel block is machined down to the final shape with more details according to the blueprint.

Nacreous shells and bones are two typical examples of natural materials with a hierarchical structure. In the nanoscale, aragonite nanograins, a natural calcium carbonate crystal and a biopolymer complex comprise the building blocks of the red abalone nacre hierarchy (Figure 6.8a). At one step further, the nano-5 grains form ~0.5 nm thick and ~6 nm wide polygon tablets surrounded by ~30 nm organic glue (biopolymers). These tablets are further organized in a staggered arrangement at the next level of the hierarchy. This type of arrangement gives a brick-and-mortar microstructure to the mother-of-pearl when viewed from a cross-section under a scanning electron microscope. At the millimeter scale, a ~20 μm thick organic layer called a growth line separates ~300 μm thick mesolayers which are visible to the eye. Finally, on the largest scale, the red abalone shell consists of two distinct layers: Calcite, the outer layer of mollusk shell made of large calcite crystals, and the inner nacre layer. The hierarchical structure of the bone is schematically presented in (Figure 6.8b) as

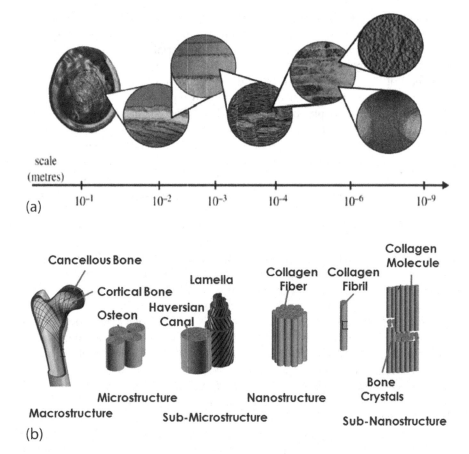

FIGURE 6.8   Hierarchical structures of (a) bone and (b) red abalone nacre.

another typical example of natural hierarchical structures. The basic building blocks of the hierarchy are nanoscale, self-assembled triple-helix collagen molecules, and plate-shaped hydroxyapatite nanocrystals. The staggered arrangement of the building blocks on a larger scale forms collagen fibers that can be organized into collagen fibers. Collagen fibers can be further arranged into lamellar bones at microstructural length scales. Finally, on the largest scale, this lamellar bone may form a cylindrical helical structure called osteon.

Finally, the nature, function, and mechanical properties of the structure are intimately interlinked in such a way that one cannot be considered exclusively without the other. For extreme situations, the function determines the structure, such as bone remodeling. This is again in comparison to the research methods used in engineering: Mechanical engineers define the function of a component and select existing materials that meet the requirements; material scientists are concerned with developing new materials or enhancing the properties of existing materials, without necessarily taking into consideration specific applications and functions for these materials.

6.5.3 BIOMIMETICS

The natural world around us offers excellent examples of usable structures consisting of a range of materials. Over the centuries, nature has evolved to adapt and develop highly sophisticated methods for solving problems. Various examples of functional surfaces, fibrous structures, structural colors, self-healing, and thermal insulation, etc., offer important lessons for future fiber products. Biomimetic research is a rapidly growing field, and its true potential for developing new and sustainable fibers can only be realized through interdisciplinary research based on a holistic understanding of nature. The design of the material mimics the structural concepts of the dermis of sea cucumbers. These creatures feature soft connective tissue with mutable mechanical properties; within seconds, the animal can switch between low and high stiffness. Nature relies on a nanocomposite in which collagen fibrils reinforce a low-modulus matrix to realize this effect. The degree of binding and stress transfer between adjacent fibrils to regulate the macroscopic properties of the system varies with the regulatory protein. Animals, plants, and insects in nature have evolved over billions of years to develop more efficient solutions, such as superhydrophobicity, self-cleaning, self-repair, energy conservation, drag reduction, dry adhesion, adaptive growth, and so on, than comparable man-made solutions to date. Some of these approaches may have motivated humans to achieve outstanding results. For example, the concept of fishing nets may have originated from spider webs; the strength and stiffness of the hexagonal honeycomb may have contributed to its acceptance for use in lightweight aircraft structures and in many other applications. Although the science of biomimetics has gained popularity relatively recently, the idea has been around for thousands of years. Since the Chinese tried to make artificial silk more than 3,000 years ago, there have been many examples of people learning from nature to develop new materials and devices. Leonardo da Vinci, for example, designed ships and aircraft by looking at fish and birds, respectively. The Wright brothers designed a successful aircraft only after realizing that birds do not flap their wings continuously; rather, they glide on air currents. In 1866, while visiting the dissecting room of the anatomist Hermann Von Meyer, engineer Karl Culmann found a remarkable resemblance between the stress lines (tension and strain lines) in the crane head and the anatomical structure of the bone trabeculae in the head of the human femur. In other words, nature reinforced the bone precisely in the manner determined by modern engineering. One of the most well-known examples of biomimetics is possibly a textile product. According to the story of George de Mestral, the Swiss inventor, he and his dog went to the woods. Upon his return, he noticed the burrs stuck to his pants and his dog's fur. Upon closer inspection of the burrs, de Mestral discovered their hook-like construction, which led to the invention of the Velcro hook and loop fastener.

There are many more examples of inventions that draw their inspiration from biological systems. This study examines the area of biomimetics as it relates to fiber. The exploration begins with a general overview, followed by a historical perspective; it outlines some ongoing efforts in biomimetic materials. Finally, it explores the potential use of biomimetic materials and products to achieve sustainable fibers. In general, the aim is not to emulate a particular biological architecture or system, but to use that knowledge as a source of guiding principles and ideas. The underlying philosophy is therefore based on what might be termed "soft interpretation," along with a large element of imagination. By combining known technologies such as lithography or surface probe microscopy. Complex functions or recognition capabilities of biological systems, micro and nanometer-scale architectures that integrate features such as anisotropy specific binding or motion have been developed for potential applications in active nano-devices dealing with electronic information and mechanical tasks, pre-coded surface coatings for clinical testing and screening, structure-function elucidation and new interface probes.

However, the interest here is in how these organic architectures can be combined with inorganic solids to produce unique and exquisite bio-minerals, e.g. diatoms frustules, coccoliths, sea shells, bones, etc., in which the composition, size, shape, orientation, texture, and alignment of the mineral constituents are precisely controlled. Bio-inorganic materials are a source of inspiration for material synthesis for many scientists. In fact, what really captured the imagination is how this relatively simple inorganic material, such as $CaCO_3$, SiO_2, Fe_3SO_4, etc., can be formed into precise functional architectures, including tough, durable, and adaptive polymer-ceramic composites, which can be produced using calcium phosphate and calcium carbonate by organized assembly based on specific molecular interactions. If these biological archetypes can be reformulated in a synthetic sense, the biomimetic design of nano-microscale materials and composites based on inorganic materials could be a real possibility in future processing strategies.

The main thrust in the field of biomimetic/bio-inspired materials is the discovery of a radically new approach to the design of bio-inspired, synthetic polymers with stimulus-responsive mechanical properties. Materials that induce a reversible change in the mechanical properties of the external stimulus are the target. This technology can also be used in a range of material systems that have the potential to enable applications ranging from biomedical implants to robotic components to the adaptation of protective clothing to orthopedic devices with controllable characteristics.

6.5.3.1 Definition of Biomimetic Material

Biomimetic materials are materials developed using inspiration from nature and the definitions of Biomimetic materials are given as follows.

(a) **Biomimetic materials** are materials invented using inspiration from nature. This may be useful when designing composite materials or material structures. Many inspiring examples have evolved from natural structures that have been used by man. Popular examples are the honeycomb structure of the beehive, the fiber structure of wood, spider silk, mother-of-pearl, bone, and hedgehog quills.

(b) **Biomimetic** is an examination of nature, its models, systems, processes, and elements to be emulated or inspired in order to solve human problems. Related terms shall include bionics. The term "biomimicry" or imitation of nature has been defined as "copying or adaptation or derivation from biology." The term "bionics" was first coined in 1960 by Steele as "a study of systems having some kind of feature copied from nature or reflecting the features of natural systems or their analogues." The term "biomimetics" introduced by Schmitt is derived from bios, meaning life (Greek) and mimesis, meaning imitation. This "modern" research is based on the belief that nature follows the path of least resistance (lowest energy expenditure) while often using the most common materials to accomplish a task. Biomimetics, ideally, should be a process of incorporating principles that promote sustainability in the same way that nature does, from "cradle to grave," from raw material use to recyclability, all in this physical world. Biomimetic material chemistry is also referred to as bio-inspiring chemistry, which is an important and diverse field, such as bio-ceramic, bio-sensing, biomedical engineering, bio-nanotechnology, and bio-induced self-assembly.

Throughout the history of life on earth, nature has gone through a trial and error process to refine the living organisms, processes, and materials on planet Earth. The emerging field of biomimetics has emerged as new technologies based on biologically inspired engineering at both macro and nanoscale levels. Biomimetics is not a new idea. Humans have been looking at nature for answers to both complex and simple problems throughout our lives. Nature has solved many of today's engineering problems, such as hydrophobicity, wind resistance, self-assembly, and the use of selective advantages for solar energy through evolutionary mechanics.

6.5.3.2 History of Biomimetic Materials

One of the early examples of biomimetics was just the study of birds to enable human flight. Although he never succeeded in creating a flying machine, Leonardo da Vinci (1452–1519) was a keen observer of the anatomy and flight of birds, and made various notes and sketches on his observations as well as sketches of various flying machines. The Wright brothers, who succeeded in flying the first heavier-than-air aircraft in 1903, were inspired by observations of pigeons in flight. Otto Schmitt, an American scholar and scientist, coined the term biomimetics to explain the transition of ideas from biology to technology. The term biomimetics only entered Websters Dictionary in 1974 and is defined as

> a study of the formation, structure or function of biologically produced substances and materials (as enzymes or silk) and biological mechanisms and processes (as protein synthesis or photosynthesis) especially for the purpose of synthesizing similar products by artificial mechanisms that mimic natural ones.

In 1960, the word bionics was coined by psychiatrist and engineer Jack Steele to describe "cell sciences that have some sort of feature copied from nature." In 1960, bionics entered Webster dictionary as "a discipline concerned with the application of data on the functioning of biological systems to solve engineering problems." The term bionic took on a different connotation when Martin Caidin referred to Jack Steele and his work in the novel "Cyborg" which later resulted in the 1974 television series "The 6 million Dollar Man" and its spin-offs. The term bionic was then associated with "the use of electronically operated artificial body parts" and "the use of ordinary human powers increased by or as if by means of such devices." The term bionic has assumed the influence of supernatural strength, and the scientific community in English-speaking countries has largely abandoned it. The term biomimicry was introduced as early as 1982. The term biomimicry was popularized by the scientist and author Janine Benyus in her 1997 book *Biomimicry: Innovation Inspired by Nature*. Biomimicry is described in her book as "a new science that studies the models of nature and then imitates or draws inspiration from these designs and processes to solve human problems." Benyus suggests that Nature is a "Model, Measure, and Mentor" and emphasizes sustainability as a biomicrobial objective. Biological material science is evolving rapidly as a new area at the intersection between material science and biology. The causes are very different. First, progress in regenerative medicine generates an ever-increasing need for new types of biomaterials with specific and well-defined interactions with the biological host system. Second, recent advances in material characterization and manufacturing technologies have prompted scientists to ask how the structure of natural materials developed in the course of evolution can be reformulated into biomimetic designs for engineering applications. This leads to new types of advanced materials with an odd combination of properties that self-assemble, repair, or grow. Finally, it is increasingly recognized that the material properties can be critical to the biological function of molecules, tissues, and organs. Therefore, the approach to material science also applies to certain fields of biology. Three separate paths constitute the vast and increasing field of biological material science (see Figure 6.9):

(a) Material science methods are used to elucidate the relationship of structure-property in biological materials. Light and electron microscopic images show the cellular network of osteocytes and the mineralized extracellular matrix surrounding them.

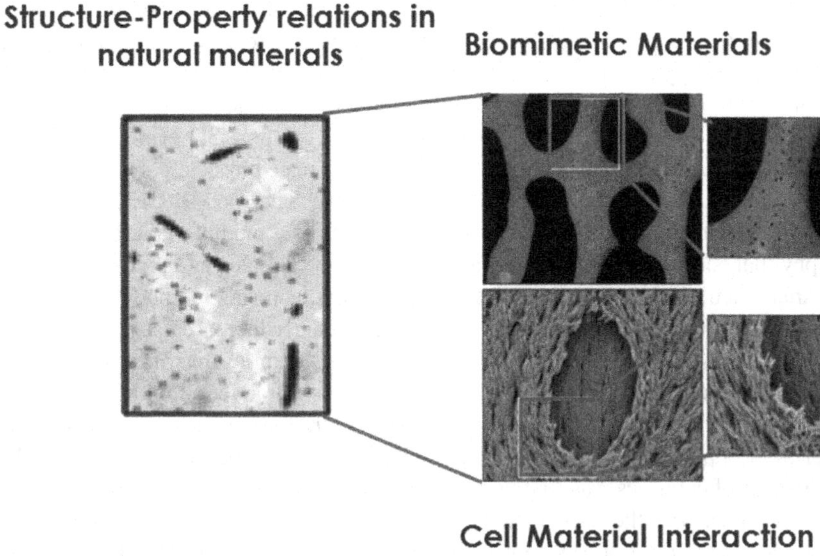

Structure-Property relations in natural materials

Biomimetic Materials

Cell Material Interaction

FIGURE 6.9 Three research areas of biological material science exemplified by research on bone.

(b) A better understanding of cell–material interactions will improve, for example, implantable biomaterials and tissue engineering. The picture shows the bone-forming cells developing on the ceramic scaffold.

(c) The design principles of natural materials may help in the development of new materials with unusual properties. The image shows a biomimetic scaffold built by rapid prototyping.

6.6 CERAMICS AND BIO-CERAMICS

6.6.1 CERAMICS

Due to its various excellent properties, ceramics are used in a wide range of applications, including the chemical industry, machinery, electronics, aerospace, and biomedical engineering. The properties that make these flexible materials include high mechanical strength and hardness, good thermal and chemical stability, and viable thermal, optical, electrical, and magnetic performance. Ceramic components are generally formed in the desired shapes, starting with a mixture of powder with or without binders and other additives, using conventional techniques, including injection molding, die pressing, tape casting, gel casting, etc. In order to achieve densification, the sintering of green parts at higher temperatures is further required. However, these ceramic forming techniques lead directly to restrictions in terms of long processing times and high costs. Structures with highly complex geometries and intertwined holes cannot be created as molding is usually part of these techniques. On the other hand, the machining of ceramic components tends to be extremely difficult due to their extreme hardness and brittleness. Not only are cutting tools subject to heavy wear, but defects such as cracking could also be caused in ceramic parts, not to mention the complexity of achieving good surface quality and dimensional precision. The advent of three-dimensional (3D) printing technologies, also referred to as additive manufacturing (AM), is seen as a manufacturing revolution. 3D printing is a series of advanced manufacturing technologies used to produce physical parts in a discreet point-by-point, line-by-line, or layer-by-layer additive way from 3D CAD models that are digitally cut into 2D cross-sections. 3D printing is a unique manufacturing philosophy that enables the flexible preparation of highly complex and precise structures that are difficult to implement using traditional manufacturing methods such as casting and machining. Productivity can also be significantly increased as many artifacts can be constructed on a single run. As a result, 3D printing has increased rapidly across science and engineering communities since its introduction in the 1980s.

The advent of 3D printing in the development of ceramic components provides completely new opportunities to address the problems and challenges described above. The 3D printing of ceramics was first documented by Marcus et al. and Sachs et al. in the 1990s. To date, with the latest developments in materials science and computer technology, a wide variety of 3D printing technologies have been specifically developed for ceramic manufacturing. Depending on the type of pre-processed feedstock prior to printing, these technologies can generally be classified as slurry-based, powder-based, and bulk solid-based processes, as outlined in Table 6.4. Note that slurry-based technologies use ceramic/polymer mixtures with viscosities ranging from low viscosity (approximately mPas) inks with low ceramic loads (up to 30 vol percent) to high viscosity (approximately Pas) pastes with much higher ceramic loads (up to 60 vol percent).

6.6.2 BIO-CERAMICS

Over the last few decades, many advanced biomaterials have been introduced in the biomedical field, including various ceramic materials for skeletal repair and

TABLE 6.4
Ceramic 3D Printing Technologies

Feedstock form	Ceramic 3D printing technology type	Abbreviation
Slurry-based	Stereolithography	SL
	Digital light processing	DLP
	Two-photon polymerization	TPP
	Inkjet printing	IJP
	Direct ink writing	DIW
Powder-based	Three-dimensional printing	3DP
	Selective laser sintering	SLS
	Selective laser melting	SLM
Bulk solid-based	Laminated object manufacturing	LOM
	Fused deposition modeling	FDM

reconstruction. These materials in the field of medical implants are often referred to as bioceramics. Bioceramics are unique in nature due to their extraordinary biological and osteo-inducing properties. These materials are unique to scaffolds due to their ability to spread, self-adhesion, and differentiate and regenerate bone tissue. In addition, excellent chemical and mechanical properties, such as improved osteo-conductivity, superior wear resistance, and biocompatibility, have made it possible to replace bone reconstruction. Bioceramics can be predicted to have a future due to a rise in bone replacement surgery each year due to an increase in aging populations. The clinical significance of the design and implantation of AM ceramic scaffolding envelops an invaluable method for the rapid and reliable production of hard tissue replacement replicas of the natural bone biological context. In view of the way in which customized scaffolding can be prepared that suits the individual patient's skeletal imperfection, layer-by-layer sintering is considered a lucrative discipline for the use of ceramic bone substitutes in regenerative medicine. In addition, the use of AM ceramic scaffolds as medicine conveyance systems is becoming more attractive and important to the bioengineering environment.

6.7 SHAPE-MEMORY MATERIALS, 4D PRINTING, AND BIO-ACTIVE MATERIALS

6.7.1 SHAPE-MEMORY MATERIALS

Materials science is all about selecting the best material available to do a specific job. For example, when you design a jet engine, you choose strong materials that are extremely light and can cope with high temperatures. You could choose aluminum, titanium, or metal alloy. But what if you want to make an aircraft component that reacts in one way at low temperatures and in another way when it heats up? This is the kind of situation where you could use a shape-memory alloy, which can reshape itself automatically as the temperature changes.

Ordinary metal artifacts do not have a memory of their form. If you sit on a pair of aluminum eyeglass frames and bend them permanently (in scientific terms, "subject to plastic deformation"), it's tricky to get them back exactly as they were. You have to use your own recollection of what the frames were originally like and painstakingly twist and bend; even then, there is no assurance that the frames would look as they used to, and they can even crack completely—through fatigue—if you wiggle them back and forth too often.

Shape-memory materials act differently. They are strong, lightweight alloys (usually two or two-metal mixtures) with very special properties. They can be "programmed" to remember their original shape, so if you bend or squeeze them, you can get them back to that original shape just by heating them up again. This is called the shape-memory effect (or thermal shape-memory effect, because thermal energy causes it to happen). Some shape-memory alloys remember one shape when they're hot and another when they're cold, so if you cool them, they're going to jump into one shape, and if you heat them, they're going to "forget" the shape and flexion into another. This is known as the two-way form-memory effect. Now, what if you could make a shape-memory object that bends and twists a huge amount but still returns perfectly to its original shape, even without heating? This is an element of shape-memory called pseudo-elasticity or super-elasticity and is used in those super-bendy, virtually indestructible eyeglass frames that manufacturers say are at least ten times more flexible than steel.

Although nitinol (also called nickel-titanium, Ni-Ti) is perhaps the most known form-memory alloy, there is much more to it, including copper, zinc, and aluminum alloys (Cu-Zn-Al); copper, aluminum, and nickel (Cu-Al-Ni); iron, manganese, and silicone (Fe-Mn-Si); and quite a few others.

6.7.1.1 How Does Shape-Memory Work

The easiest way to understand shape-memory is to understand that what's going on inside a material (at the nanoscale of atoms and molecules) can be quite different from what seems to be going on outside.

Stretch the elastic band and, inside, untangle the long, knotted rubber molecules and cut them apart. Remove the stretching force and the molecules will pull back together again. That's exactly how elasticity functions. The memory of the form is different. Bend an object made of a shape-memory alloy and deform its internal crystalline structure. Let it go, and it stays as it is, bent out of shape permanently. Now apply some heat and the crystalline structure inside shifts to a completely different form, causing the object to return to its original form. Pseudo-elasticity is similar, but you do not need to change the temperature to make the object return to shape after you deform it. If you bend a pair of shape-memory eyeglasses, the stress that you apply makes the titanium alloy from which they are made flip into a completely different crystalline structure; let go, and the crystalline structure returns, so that the glasses return to their original shape.

What happens to shape-memory and pseudo-elasticity is that the internal structure of a solid material shifts back and forth between two very different crystalline forms: In other words, the molecules rearrange them in a totally reversible way. It's called a solid-state phase change—and it sounds more complicated than it actually is. We are all used to phase changes: Every time you place an ice cube in a drink and watch it melt, you are seeing a change of phase. As the frozen water heats up, its molecules change from being in a tightly packed rigid structure to an arrangement that is much looser and more fluid, so that the water changes from its solid phase (ice) to its liquid phase (ordinary liquid water). A broadly similar thing happens in a solid-state phase change, it's just that the material is solid both before and after transformation, because the molecules remain very close together.

Shape-memory alloys rotate back and forth between two solid crystalline states called austenite and martensite. At lower temperatures, they take the form of martensite, which is relatively soft, plastic, and easy to shape; at a (very specific) higher temperature, they transform into austenite, which is a tough material and much more difficult to deform. (Note that this is different from steel, where martensite is the hardest of the two.) Let's say you have a shape-memory wire and you can bend it into new shapes relatively easily. It is martensite inside, and that's why it's easy to deform. No matter how you bend the wire, it's still in its new shape; like any ordinary wire, it seems to be undergoing a very ordinary plastic deformation. But now it's the magic part! Heat it up a little, above its transformation temperature, and it will turn into austenite, with the heat energy you supply to rearrange the atoms inside and turn the wire back into its original shape. Now cool it down and it's going to return to the martensite, still in its original form. If the material is above its transition temperature all the time, you can deform it, but it will spring back to shape as soon as you release the force you apply.

The surprising (and, for some people, confusing) thing about shape-memory is that the change between austenite and martensite is not a "symmetrical" one. You can take a "programmed" piece of shape-memory wire (one that has a definite shape that you can remember) and bend it in any number of different ways. But, having done that, if you heat up your randomly bent piece of wire, it will always turn back into a single, very definite shape. We can appreciate this by understanding that our material can happily take any number of crystalline forms while it is in a martensite state. But when it's in the form of austenite, there's only a crystalline form it can take. It is the most stable form—the one with the lowest state of energy.

6.7.1.2 Shape-Memory Alloys

Despite the fact that small- and medium-sized enterprises were found in an AuCd alloy as early as 1932, the attraction of this phenomenon was not so evident until 1971, when significant retrievable strain was observed in an NiTi alloy at the Naval Ordnance Laboratories, United States. A wide variety of SMAs have now been produced in solid, film, and even foam shapes. Just three alloy systems, namely NiTi-based, Cu-based (CuAlNi and CuZnAl), and Fe-based, are currently of more commercial value. A systematic comparison of the NiTi, CuAlNi, and CuZnAl SMAs has been made with respect to the various performance indexes of interest to engineering applications. NiTi should be the first choice because it has high performance and good biocompatibility. This is crucial for biomedical applications, such as stents and guide wires in minimally invasive surgery. Cu-based SMAs have the advantages of low material cost and good workability in processing, and some of them have a rubber-like behavior after aging in a martensite state. The SME in Fe-based SMAs is generally known to be relatively much weaker and Fe-based SMAs were most likely used only as a one-time actuation fastener/clamp, largely due to the extremely low cost. However, a ferrous polycrystalline SMA has recently been reported with an enormous superelasticity (13 percent) and high tensile strength (over 1 GPa). All of these SMAs are thermo-responsive, i.e. heat is the stimulus needed to trigger the shape recovery. In recent years, good progress has been made in developing magneto-responsive ferromagnetic SMAs. However, thermo-responsive SMA has matured more from the point of view of real engineering applications, and many commercial applications have so far been realized.

With the current trend toward micro-electromechanical systems (MEMS) and even nano-electromechanical systems (NEMS), thin film SMAs (mainly NiTi-based, produced by sputter deposition) have become a promising candidate for motion generation in these micron/submicron systems. This is further supported by the finding that the SME even occurs in nano-sized SMAs and that the laser beam can be used for local annealing and/or controllable growth of SMA thin films. In addition to the SME, some SMAs also have a temperature memory effect (TME) so that the highest temperature(s) in the previous heating process(es) within the transition range can be registered and precisely detected in the next heating phase. Partially different martensite after the thermal programming phase is assumed to be the underlying mechanism for the TME. On the basis of the same principle, a piece of SMA strip can be thermo-mechanically programmed to bend forward and then backward after heating (Figure 6.10). This is a kind of phenomenon, known as the multi-SME, in which a piece of SMM recovers its original shape step-by-step through one or a few intermediate shapes. Multi-SME can be used to work virtually as a machine, but the fascinating point here is that the material is a machine. Since there is only one intermediate shape, the behavior shown in Figure 6.10 is triple-SME. It should be noted that, fundamentally, this phenomenon does not share common ground as a multi-SME in a piece of SMA with a gradient transition temperature, either by pre-straining or local heat treatment.

6.7.1.3 Shape-Memory Polymers

From the engineering point of view, the material properties of polymers are much simpler to modify than those of metals/alloys. In addition, the cost (both material

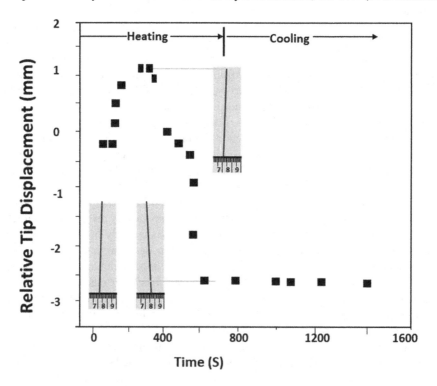

FIGURE 6.10 Triple-SME in an NiTi SMA strip upon heating. Since there is no apparent visible deflection during subsequent cooling, the material does not have the two-way SME.

and processing costs) of polymers is traditionally much lower. A number of SMPs have been discovered and well recorded in literature, while new ones are still emerging every week, if not every day. In addition to the above-mentioned advantages, the SMPs are much lighter, have a much higher (at least higher order) recoverable strain than the SMAs, and can be triggered at the same time for the recovery of shape by various stimuli and even multiple stimuli. In addition to heat, light (UV and infrared light) and chemical (moisture, solvent, and pH change) are two of these types of stimuli. In addition, many SMPs are naturally biocompatible and biodegradable. As a result, we have more degrees of freedom to manipulate SMPs to meet the needs of a specific application.

The thermoplastic polyurethane SMP was originally invented by Dr. S. Hayashi at the Mitsubishi Heavy Industry Nagoya R&D Center in Japan and has been successfully marketed for over 15 years. The same SMP was developed into open-cell foams for space missions and biomedical applications based on the concept of cold hibernated elastic memory (CHEM) proposed by Dr. W. Sokolowski at Jet Propulsion Laboratory, United States. Figure 6.11 demonstrates the possibility of delivering a coil, consisting of a thermo-moisture responsive SMP, to a jellyfish. This concept can be extended to, for example, the delivery of a tiny machine to a living cell for future cell surgery or to a tiny microbe for special function/operation within a microbe. Biomedical application is currently emerging as a promising field for SMPs, although surface patterning (to modify various surface-related properties, such as reflection, surface tension, etc.) is another area. Figure 6.12 is a zoom-in view of the patterned

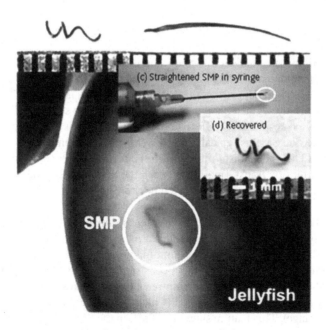

FIGURE 6.11 Delivery of an SMP coil into a jellyfish by injection. (a) Original coiled shape; (b) After being straightened at high temperature; (c) Ready for injection; (d) Recovered shape.

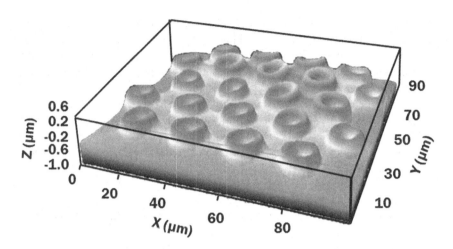

FIGURE 6.12   Zoom-in view (three-dimensional) of patterned surface atop SMP produced by laser after one single exposure.

surface above the SMP, which is produced by a laser beam after a single exposure through a microlens, compared to patterning on top of SMAs, in particular for different shaped/size patterns. Indentation-polishing-heating (IPH) patterns are made, while those at the top of the SMAs are usually reversible between two shapes during thermal cycle; those at the top of the SMPs are fixed and permanent. A further step is to integrate micro-/nano-sized wrinkles with such surface patterns for dramatically improved performance, such as self-cleaning, cell adhesion, water splitting, and light extraction, etc.

The underlying mechanism for small- and medium-sized enterprises in SMPs is the dual-segment/domain structure (one is often hard/elastic, while the other may be soft/ductile or rigid depending on whether the stimulus is correct). The former is known as the elastic segment, and the latter is the transition segment. Take an example of the thermo-responsive SMP. The SME mechanism is shown in Figure 6.13. As we can see, the SMP is typically much weaker at high temperatures than at low temperatures. This mechanism differs from the well-known and highly predictable reversible martensitic transformation between the high-temperature austenite phase (which is hard and rigid) and the low-temperature martensite phase (which is soft and flexible) in SMAs.

Since SMPs have a much greater recoverable strain and, as a rule, a wider recovery temperature range, it is possible to have more than one intermediate shape in the multi-SME through a proper programming procedure, as Xie has recently demonstrated. We have shown that during a constrained recovery (i.e. heat with a temporary SMP shape fixed), the maximum reaction force/stress should appear at the temperature that the SMP is deformed (which should be within the Xie transition temperature range). This feature shows the TME in the SMPs. The exact mechanism behind this function is the step-by-step release of the elastic energy stored in the elastic section during programming, which varies from that of the SMAs.

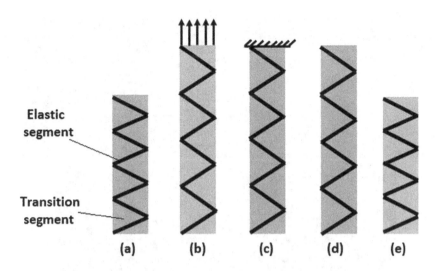

Elastic segment

Transition segment

(a) (b) (c) (d) (e)

FIGURE 6.13 Illustration of the mechanism of the SME in thermo-responsive SMP. (a) Hard at low temperature; (b) Easily deformed at high temperature; (c) Hard again after cooling; (d) Temporary (deformed) shape after constraint removed; (e) Shape recovery upon heating.

Conversely, multi-SME in SMPs can be accomplished by setting various shape recovery conditions, e.g. different stimuli or multiple changes across different temperature ranges and even a gradient transition temperature. As such, the programmed recovery can be realized in a well-controlled manner.

SMP composites have remarkably broadened the potential applications of SMPs; in addition to the use of various types of fillers (including different types of clay, SiC nanoparticles, etc.) to strengthen SMPs. Heating of thermo-responsive SMPs can also be achieved by heating Joules (by means of filling with different forms of conductive inclusions), by induction heating (by means of dissipation of energy by hysteresis by applying an alternating magnetic/electrical field, etc.), and also by radiation. Figure 6.14 shows scanning electron microscope images that show the dispersion of Ni powder (3–7 µm in diameter) in SMP/carbon black composite polyurethane. As reported by Leng with an additional 0.5 vol percent of Ni powders, and after forming chains (under a weak magnetic field before curing), the electrical resistance of SMP/carbon black composites reduces by more than ten times, so that low electrical voltage heat can be achieved, just like SMAs.

As compared to SMAs, SMPs usually soften in the presence of the right stimulus; thus, most SMPs (with the exception of a few) are not appropriate for cyclical action and cannot be conditioned to have the so-called two-way SME (which is the ability to repeatedly move between two shapes depending on how the stimulus is applied).

On the other side, the shape recovery of SMPs can be followed by color change, excellent clarity, reversible adhesion/peeling, and even self-healing. The SME has been shown to be similar to SMAs in submicron size SMP (Figure 6.15). While it has been difficult to manufacture high-quality porous SMAs so far, it is always easy to produce SMP foams using many conventional polymer foaming techniques. As

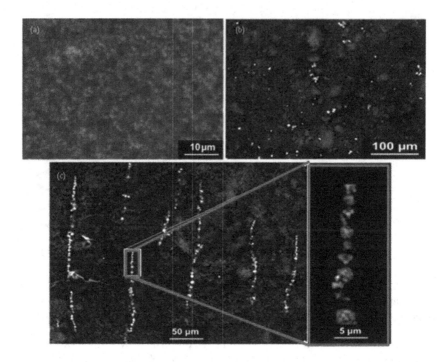

FIGURE 6.14 Enhanced electrical conductivity in an SMP/carbon black composite through alignment of Ni powder (SEM image). (a) SMP/carbon black only; (b) With randomly distributed Ni; (c) Chained Ni.

reported in the complete form recovery, 94 percent of pre-compressed polyurethane SMP foam is detected even after a few months of storage.

6.7.1.4 Shape-Memory Composites

We have seen that SMPs can be synthesized/designed to have the required properties for a specific application. However, trial and error, as well as a strong background (professional knowledge and experience), is needed. Form memory composites (SMCs), which contain at least one type of SMM, either SMA or SMP, as one of the components can be managed easily by design engineers, if the properties of SMA/SMP are well understood. Via careful design, along with the introduction of additional mechanisms (e.g. elastic buckling), further phenomena and new features can be realized in SMCs (e.g. polymer self-healing).

6.7.1.5 Shape-Memory Hybrids

Shape-memory hybrid (SMH) is a more open and versatile solution for the average user, even with minimal scientific/engineering history. SMHs are made of conventional materials (properties are well known and/or can be easily found, but all without the SME as an individual). Thus, the SMM can be designed in a do-it-yourself (DIY) manner to achieve the required function(s) in a specific application.

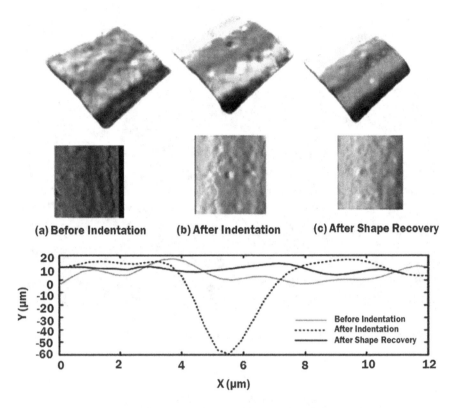

FIGURE 6.15 The SME demonstrated by indentation test (using a Berkovich diamond indenter) in a 170 nm thick thin film polyurethane SMP. Top: Three-dimensional and two-dimensional surface scanning images. Bottom: Cross-sectional view.

Similar to SMPs, SMHs are also based on a dual-domain system in which one is always elastic (elastic domain) while the other (transition domain) is able to change its stiffness remarkably if a correct stimulus is presented. However, the selection of transition domains for SMHs must be based on the principle that any possible chemical interaction between the elastic domain and the transition domain should be reduced if it cannot be entirely avoided. This is even better than an organic-inorganic mix. As such, we can accurately predict the thermo-mechanical response (or any other of our concerns) of the SMH based on the material properties of these two domains. The benefits of the SMHs are evident. For example, the elastic domain can be selected to meet the requirements for the stiffness and the sum of the shape recovery ratio of the SMH, while the appropriate stimulus form can be realized by selecting the correct material for the transition domain. In addition, the manufacturing/synthesis method is based on the well-known properties of domain materials, which is simple. It is manageable by almost anyone, even without a lot of experience.

We have developed a silicone-based SMH system for demonstration purposes. All features found in conventional SMAs and SMPs, namely dual-SME, triple-SME (as shown in Figure 6.16), two-way reversible SME (as shown in Figure 6.17), thermo-responsive (including by means of Joule heating by passing electrical current),

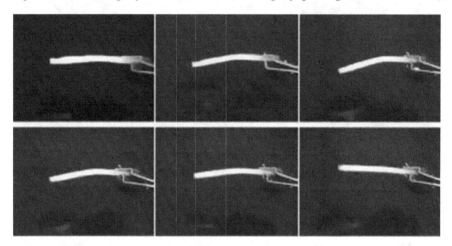

FIGURE 6.16 Triple-SME. Upon immersing into hot water, the SMH beam bends downward and then upward.

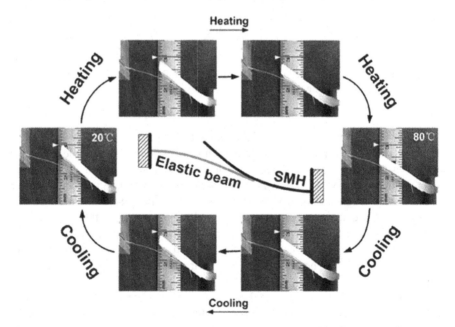

FIGURE 6.17 Two-way SME. Upon heating (top half, from left to right), the SMH beam pushes down the elastic beam (see inset for the illustration of set-up). Subsequently, upon cooling (bottom half, from right to left), the elastic beam pushes the SMH beam upward.

thermo/moisture-responsive, etc., have been reproduced. A narrow temperature recovery range within 5°C has been achieved. This principle has been further applied to the design of pressure-responsive SMH, thermo (on cooling or at extremely high temperatures)-responsive SMH with some success.

In addition, SMH has been shown to be rubber-like (not only in the high temperature range as superelastic SMAs, but also within the full application temperature range both below and above the transition/recovery temperature and, more importantly, with small hysteresis, as shown in Figure 6.18). Irrespective of the rubber-like feature, after cooling, the SMH retains more than 90% of the amount of pre-strain when extended to double its length at high temperatures. Subsequently, full-form recovery is observed immediately after heating in hot water. More interestingly, this SMH has a real crack healing function when it's heated, but not just for shape recovery and crack closure. Unlike conventional polymer self-healing, which involves a curing agent for polymerization and can only be used once, after more than ten cracking/healing cycles, this SMH does not display any deterioration in terms of the strength recovered after healing.

For a more successful demonstration of the repeatable self-healing function, we manufactured a piece of cylindrical SMH with a piece of SMA spring embedded inside, as illustrated in Figure 6.19a. Remember that this SMA spring can be easily extended in a seemingly plastic way at room temperature. Upon heating, it returns to its original shape/length. We pulled the SMH until it broke into two pieces, as shown in Figure 6.19b1. The SMA spring was also extended, resulting in a wide gap between the two sections of the SMH even after the pulling force was completely

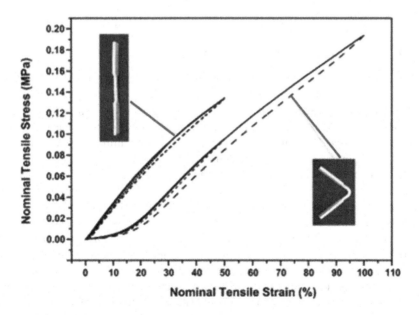

FIGURE 6.18 Rubber-like SMH (in original and pre-bent shapes) under cyclic loading at room temperature (about 22°C) and at high loading speed.

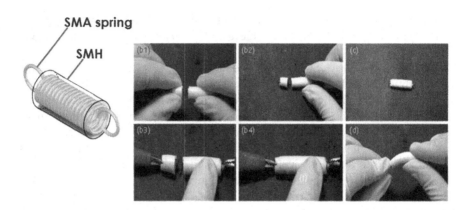

FIGURE 6.19 Self-healing of SMH. (a) SMH with an embedded SMA spring inside (illustration); (b1-b2) Pulling SMH till fracture; (b3-b4) Joule heating SMA coil for self-healing; (c) Healed sample; (d) Healed sample in bending.

removed (Figure 6.19b2). Subsequently, an electrical current was passed through the SMA spring (Figure 6.19b3). The SMA was heated for the recovery of the form so that the gap was closed (Figure 6.19b4). In the meantime, the SMH was also heated to achieve self-healing, with the aid of the compressive force produced by the SMA spring during yoke heating. After cooling back to room temperature, the SMH became one piece again (Figure 6.19c). There was no apparent crack even when it was severely bent, as shown in Figure 6.19d. This cracking-healing procedure can be repeated several times.

6.7.2   4D PRINTING AND BIO-ACTIVE MATERIALS

One of the primary objectives of 3D printing in health science is to mimic biological functions. To accomplish this goal, 4D printing can be applied to 3D-printed objects that will be distinguished by their ability to change over time and under external pressure by changing their form, properties, or composition. Such capabilities are a promise of great opportunities for biosensing and biomimetic systems to move toward more physiological imitation systems.

Bioactive material is a material capable of interacting actively with biological compounds or living organisms. This ability typically originates from the material-integrated recognition components, which are biomolecules (enzymes, antibodies, carbohydrates, living cells, etc.), chemical functions (adhesive or repulsive surfaces), or surface microstructures. Active interactions can lead to sensing (in the biosensor field) or mimicking behavior (in the tissue engineering field) depending on the field of application of the bioactive material. When dealing with sensing material, the interaction occurs with a particular analysis and produces a signal that is transformed into a measurable response. When dealing with tissue engineering material, the interaction can be unique to the tissue being studied and can contribute to biomimetic activity (elasticity, cell growth, strength, movement, etc.).

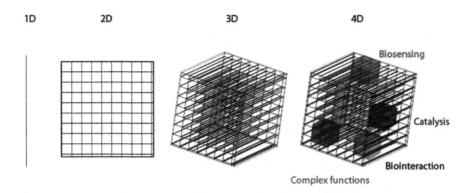

FIGURE 6.20   From 1D to 4D concept: 4D printing is adding new capabilities to multi-material 3D-printed objects: Either physical, electronic, chemical, biological, or biochemical properties.

The 4D concept was first introduced into the architectural lexicon in 2013 by Skylar Tibbits, founder of the Self-Assembly Lab hosted by the Massachusetts Institute of Technology (MIT) International Design Center. This idea introduces the notion of time in 3D printing to add new capabilities to multi-material 3D objects. For Skylar Tibbits and his team, as well as several other teams involved in the field, 4D materials are constructed through 3D printing in such a way that they later react and change shape in response to external stimulus such as heat or moisture. One example of such a well-known system is their Hilbert Gen 1 Water self-folding tool. Similar work was done by Jennifer Lewis' team at the Wyss Institute for Biologically Inspired Engineering at Harvard University, which proposed several flower-shaped devices inspired by natural structures that respond to and alter their shape in response to environmental stimuli.

The definition of 4D is generalized so as not to limit 4D printing to the ability to program physical and biological materials to change shape, but also to describe 4D printing as adding new capabilities to multi-material 3D printing tools: Either physical, electronic, chemical, biological, or biochemical properties (Figure 6.20). This concept is now also complemented by the definition of bioprinting, which is now widely classified in this 4D printing category, due to its post-printing maturation steps that occur over time.

6.8   ADVANCED AM MATERIALS

The main objective of additive manufacturing technology is to allow the development of high-performance, more advanced parts and components. The creation of new and more sophisticated materials is therefore a crucial factor in the advancement of AM technologies. Developing and refining advanced 3D printing materials allows faster processing and more complex parts to be created; at the same time, the advancement of 3D printing processes depends on refining ever more advanced materials for higher-performance parts. In general, by means of advanced materials, the industry refers to all those materials which are at the cutting edge of material science production. These include composites, high-performance polymers, high-temperature metals (refractory), and ceramics.

Composites used in AM are mainly composites of carbon fiber, glass fiber, or Kevlar inside a thermoplastic matrix (unlike conventional composites in a thermoplastic matrix). They can be made available as powders, pellets, or filaments and typically used as chopped fibers (although technologies capable of producing continuous fiber composites are in development). Nevertheless, the concept of composites is incredibly broad and can be expanded to include metal composites, polymer-ceramic composites, and even carbon-ceramic composites.

Advanced AM materials also include high-performance polymers, in particular PAEK family polymers such as PEEK, PEKK, and PEI (ULTEM). These are high-temperature, high-stress thermoplastics that melt at temperatures of 400°C. The ability to print 3D using these materials for a number of medical and metal-replacement industrial applications is expected to significantly boost the demand for additive manufacturing.

Ceramics is also a very large family of materials that can range from cement and clay to diamonds. Advanced 3D ceramic printing materials are primarily represented by technical ceramics such as alumina (aluminum oxide), zirconia (zirconium oxide), and other advanced ceramic materials based on silicon. These materials offer almost unparalleled properties in terms of heat resistance, strength, and light weight, but they are difficult to shape using traditional technologies. This is why they are considered especially important for the future of AM.

Advanced materials in metal additive manufacturing include mainly refractory metals (as well as certain metal composites). These are extremely high-temperature resistant metals that can withstand temperatures in excess of 4000°C. The list of 3D printable refractory metals (mainly powder-bed fusion processes) includes tungsten, niobium molybdenum, and other metals that can greatly benefit from the geometrical possibilities provided by AM. Finally, 3D printing soon became a critical technology for the creation of completely new categories of materials. Which involve applications involving graphene and graphene allotropes (the different forms that graphene may produce, such as nanotubes or buckyballs), ultra-light or ultra-dark or even so-called 4D materials.

Broadly speaking, these are materials that can be programmed to perform a particular action after they have been used for the manufacture of an object. They therefore add the "4th dimension" in that the objects continue to evolve over time after the 3D printing process has been completed. For example, shape-memory materials that can turn back to previous states and shapes, or auto-assemble structures that change shape as they heat up or cool down. The list goes on and on, and in many cases 3D printing has proved to be of key importance in allowing rapid testing of new theories in material behavior.

6.9   SUPPORT MATERIALS

In FDM 3D printing, support structures are required when printing has overhangs or features suspended in mid-air. They allow the successful printing of complex shapes by protecting these otherwise unsupported areas.

The support material for 3D printers is simply the material in which these supports are printed. Different materials provide different compromises between price, ease of use, and print quality, so selecting the right material for your models can make your printing experience much more enjoyable.

6.9.1 BUILD MATERIAL SUPPORTS

The easiest, most common support material is essentially the same material from which model is made. This is because many 3D printers are still single extruder machines capable of printing only one material at a time, making a dedicated support material impossible (Figure 6.21).

In addition to being more accessible, common building materials tend to be more affordable than dedicated support materials. Build material supports may therefore be an attractive option for those on a budget. Build material Installed structure supports are also expected to conform better to the pattern, made of the same structure. However, this adhesion is a double-edged sword: Prints are less likely to fail, but support for removal, will require more effort and will yield poorer surface quality. As shown in Figure 6.22, the use of an X-Acto knife or sandpaper may be required to achieve a smooth surface on the model after removal of the support.

6.9.1.1 Materials as Their Own Support

Since PLA is more fragile, if your model requires smaller support clusters that are less dense, it may be more difficult to remove than ABS and PETG, which are more ductile materials. Besides that, anecdotal evidence indicates that the printing of PLA supports under PETG, and vice versa, has good results because the two materials do not bind together and can therefore be clipped more easily.

Works on: Single extruder, multi-extruder 3D printers
Pros:

- Works on single extruder 3D printers
- More competitive
- Compatibility of materials is not a problem

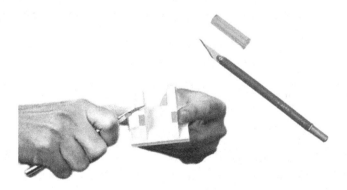

FIGURE 6.21 Cleaning up build material supports with an X-Acto knife.

FIGURE 6.22   Breakaway supports being removed.

Cons:

- Poorer consistency of surface
- Removal of support may be a problem

6.9.2   QUICK REMOVAL: BREAKAWAY SUPPORTS

Breakaway promotes research much like construction materials. Yet since breakaway supports have a different content, they are less likely to over-adhere to your prints. It means that the same support systems printed in breakaway materials would be much easier to remove and leave much cleaner surfaces behind (as shown in Figure 6.22).

Material compatibility is a problem, as is often the case with multi-material 3D printing, because the supporting material cannot conform adequately to all building materials. The quality of breakaway support materials is also low, which is the least popular choice on this list. Options include Matterhacker's PRO Series Breakaway, Ultimaker's Breakaway, E3D's Scaffold (also water-soluble), and Scaffold's Snap (only breakaway) materials.

Works on: 3D multi-extruder printers
Pros:

- Fast and clean removal

Cons:

- Compatibility of materials is a factor
- Low availability
- Works only on multi-extruder 3D printers

6.9.3 BEST QUALITY: SOLUBLE SUPPORTS

Soluble support materials offer great possible surface quality. Instead of requiring manual removal, soluble supports simply dissolve away, leaving behind clean, unmarred surfaces. This allows for dense, solid supports with optimum dimensional accuracy, as well as complex geometries to which solvents, but not pins or X-Acto blades, can be applied. Moving assemblies, such as the gyro above, also benefit from such dense support and clean removal (Figure 6.23).

Apart from being literally more expensive, soluble support materials are usually harder to monitor. The most common water-soluble material, PVA, is extremely hygroscopic and therefore degrades with moisture. Proper handling of the filament is a must. Material compatibility is also a concern, as no single soluble material works well with all building materials. If PVA doesn't work, alternatives such as HIPS will often require nasty chemicals to dissolve, which is less than ideal. Soluble support can take up to two hours to completely dissolve, making it a poor option for those pressed for time.

Works on: Multi-extruder 3D printers
Pros:

- Cleanest removal and surface finish
- Complex structures can be supported (internal structures, moving parts)

Cons:

- Material compatibility is a consideration
- Material storage can be a hassle
- Dissolving times are long
- Only works on multi-extruder 3D printers

FIGURE 6.23   Complex geometries are no match for dissolvable support material!

6.9.4   SUPPORTING WITH THE ORIGINAL MATERIAL

This is the only option when you print with one extruder. Using this process, the necessary support is printed with the same material as the product. This method is simple to use; you only need one material. Such support structures can be generated by a slicing software package, such as simplify 3D. Note that it is important not to use too much supporting material, as supporting structures of the same material are more difficult to remove from the model than other options.

6.9.5   SUPPORTING WITH PVA FILAMENT

There are various special support filaments available that are completely soluble, and PVA is one of them. You need a 3D printer with a double extruder to be able to print with PVA. PVA stands for polyvinyl alcohol and is a soft, biodegradable polymer that is very prone to moisture. As PVA is exposed to sunlight, it dissolves. This is also ideal as a support material for 3D printing. After printing, the filament can be quickly extracted by dissolving it in cold or warm water. PVA is still used in conjunction with PLA filament, but is now also added more and more to other filaments such as PET-G. In addition, there are several new modifications that make it possible to use PVA with higher temperatures.

6.9.6   SUPPORTING WITH PVA+ FILAMENT

Previously, HIPS was mostly used as support material for ABS printing. With the arrival of PVA+, HIPS is used much less. The explanation for this move is that HIPS must be dissolved in limonene. It is a radioactive material that is difficult to acquire. HIPS is also often substituted by PVA+ (modified PVA), a filament that is easily soluble in water, much like PVA. At this time, PVA+ is being tested for its suitability to be combined with all other filaments. It is also necessary to use a double extruder for PVA+.

The major advantage of printing with support material is that it is quickly removed without leaving pieces behind or damaging the 3D model. The disadvantage is that support filaments are often more expensive than base filaments and can only be printed on a 3D printer with a double extruder.

6.10   EXERCISES

1. Explain the factors to be considered when choosing materials for additive manufacturing.
2. Describe the multi-material and different composite additive manufacturing methods.
3. What are the different metal AM processes and materials? Explain in detail.
4. Explain the composite materials in details.
5. Discuss biomaterials, hierarchical materials, and biomimetics.

6. Explain ceramics and bio-ceramics materials.
7. Briefly describe the history of biomimetic materials.
8. Explain the concept of shape-memory materials, 4D printing, and bio-active materials.
9. State and explain the different advanced AM materials.
10. Explain support materials.

6.11 MULTIPLE-CHOICE QUESTIONS

1. When choosing a material and 3D printing method for your project, make sure your material suits the _____ needed for the application.
   a) Certifications and/or key features
   b) Color and appearance
   c) Quality
   d) None of the above
   Ans: (a)

2. ABS materials have a _____ melting point.
   a) Low
   b) Moderate
   c) High
   d) Very low
   Ans: (c)

3. Stainless steel is printed by fusion or _____.
   a) Stereolithography
   b) Laser sintering
   c) Material jetting
   d) None of the above
   Ans: (b)

4. Stereolithography (SLA) has been developed on the basis of _____.
   a) Photopolymerization phenomena
   b) Material jetting
   c) Drop on demand
   d) None of the above
   Ans: (a)

5. The sintering temperature of the printed ceramic component can affect the _____ fraction of the metal phase.
   a) Viscosity and gravity
   b) Thickness and length
   c) Density and volume
   d) All of the above
   Ans: (c)

6. The aluminum content is a combination of _____ and a very low amount of gray aluminum powder.
   a) Polylactide
   b) Ceramic
   c) Aluminum
   d) Polyamide
   Ans: (d)

7. The 3D printing of ceramics was first documented by Marcus et al. and Sachs et al. in the _____.
   a) 1990s
   b) 1980s
   c) 1970s
   d) 1960s
   Ans: (a)

8. The 4D concept was first introduced into the architectural lexicon by Skylar Tibbits in _____.
   a) 2005
   b) 2008
   c) 2009
   d) 2013
   Ans: (d)

7 Applications and Examples

7.1 APPLICATION–MATERIAL RELATIONSHIP

When it comes to 3D printing, the sky is (almost) the limit in terms of what materials you can use—and researchers are constantly making new 3D printable materials.

There are a few major types of materials used in 3D printing. Also popular are plastics that can vary from engineering grades, such as PEEK, to very easy-to-use plastics, such as PLA. Resin is another common material used with SLA printers. Composites are another type and as the name suggests, they are generated by combining two materials in order to obtain the best properties of product. Metals are the last major category of materials, which can only be printed using modern equipment.

7.1.1 POLYMER

From realistic display models to functional prototypes, tooling, and end-use parts, the opportunities created by polymer 3D printing are immense. Did you know that some high-performance thermoplastics are even tougher than aluminum?

a) Polyamide (PA 6)
   Method: Powder bed fusion
   Technology: SLS, MJF
   Applications: Piping and media flow/storage parts; Fluid reservoirs; Multi-purpose industrial goods
   Tensile strength (MPa): Ranging from 38 to 66
   Elongation at break (%): Ranging from 1.6 to 16
   Hardness: Object, print direction, and technology dependent
b) Polyamide (PA 11)
   Method: Powder bed fusion
   Technology: SLS, MJF
   Applications: Insoles; Living hinges; Prostheses; Snap fits
   Tensile strength (MPa): ~50
   Elongation at break (%): Ranging from 35 to 50
   Hardness: ~Shore D 80
c) Polyamide (PA 12)
   Method: Powder bed fusion
   Technology: SLS, MJF
   Applications: Connectors; Drones; Enclosures; Housings
   Tensile strength (MPa): Ranging from 41 to 48

Elongation at break (%): Ranging from 15 to 20
Hardness: ~Shore D 80

d) Glass Bead Filled Polyamide
Method: Powder bed fusion
Technology: MJF
Applications: Fixtures; Tooling; Enclosures; Housings
Tensile strength (MPa): ~30
Elongation at break (%): ~10
Hardness: ~Shore D 82

e) TPU
Method: Extrusion; Powder bed fusion
Technology: FFF/FDM, SLS, MJF
Applications: Footwear, hoses, and tubes; Sealings
Tensile strength (MPa): ~10
Elongation at Break (%): ~250
Hardness: ~Shore A 90

f) Silicone
Method: Vat polymerization; Material jetting
Technology: DLP, SLA, DOD
Applications: Sealings, molds, medical devices
Tensile strength (MPa): Ranging from 6 to 9
Elongation at break (%): Ranging from 160 to 800
Hardness: Ranging from Shore A 20 to 70

g) PEEK
Method: Extrusion; Powder bed fusion
Technology: FFF/FDM, SLS
Applications: Aerospace, medical, electrical
Tensile strength (MPa): Ranging from 85 to 100
Elongation at break (%): Ranging from 2.6 to 3
Hardness: Object and technology dependent

h) PEI
Method: Extrusion
Technology: FFF/FDM
Applications: Aerospace, automotive, electrical
Tensile strength (MPa): Ranging from 70 to 80
Elongation at break (%): Ranging from 3 to 6
Hardness: Object and technology dependent

i) Polypropylene
Method: Extrusion; Powder bed fusion
Technology: FFF/FDM, SLS
Applications: Low-friction mechanical parts and food packaging
Tensile strength (MPa): Ranging from 20 to 25
Elongation at break (%): Ranging from 20 to 75
Hardness: ~Shore D 65

j) Polycarbonate
  Method: Extrusion
  Technology: FFF/FDM
  Applications: Brackets, fixtures, clamps
  Tensile strength (MPa): ~57
  Elongation at break (%): ~4.8
  Hardness: ~Rockwell R 115
k) ABS
  Method: Extrusion
  Technology: FFF/FDM
  Applications: Models, alignment jigs, light prototyping
  Tensile strength (MPa): Ranging from 33 to 41
  Elongation at break (%): Ranging from 4 to 6
  Hardness: ~Shore D 109

7.1.2 METAL

Some of the most difficult 3D printing materials are metals. They also deliver thermal properties and high strength. Most of those metals are suitable for various applications in several alloys.

a) Stainless Steel
  Method: Extrusion; Binder jetting
  Technology: FFF/FDM, BJ
  Applications: Tools, gears, jewelry, miniatures, molds
  Tensile strength (MPa): Ranging from 521 to 582
  Elongation at break (%): Ranging from 36 to 55
  Hardness: ~Rockwell B 71
b) Aluminum
  Method: Powder bed fusion
  Technology: SLM, DMLS
  Applications: Spare parts, functional components
  Tensile strength (MPa): Ranging from 410 to 440
  Elongation at break (%): Ranging from 4 to 6
  Hardness: ~Brinell HB 119
c) Titanium
  Method: Powder bed fusion
  Technology: SLM, DMLS
  Applications: Biomedical implants and tooling, jewelry
  Tensile strength (MPa): Ranging from 1000 to 1200
  Elongation at break (%): Ranging from 7 to 11
  Hardness: ~Rockwell B 40
d) Maraging Steel
  Method: Powder bed fusion
  Technology: SLM

Applications: Furnace parts; Tooling
Tensile strength (MPa): ~1135
Elongation at break (%): ~11
Hardness: ~Vickers HV10 373

e) Cobalt-Chrome
Method: Powder bed fusion
Technology: SLM, DMLS
Applications: Engine parts; Furnace parts; Implants
Tensile strength (MPa): Ranging from 1050 to 1450
Elongation at break (%): Ranging from 8 to 28
Hardness: ~Rockwell C HRC: 35

f) Tungsten
Method: Powder bed fusion
Technology: SLM
Applications: Balance weights, MRI
Tensile strength (MPa): N/A
Elongation at break (%): N/A
Hardness: Ranging from Vickers HV30 300 to 650

7.2 FINISHING PROCESSES

3D printing is a revolutionary emerging technology that enables manufacturers to produce components fairly rapidly, in small batches, and with a high degree of flexibility. Sometimes, though, when a component comes out of a printer, it needs to undergo a surface finishing process before it's ready for use, particularly if it's a consumer product or a plastic piece you're going to use under harsh conditions.

For several of the same purposes, you can want to finish a 3D-printed component by applying finishing processes to a product created using injection molding or other manufacturing technology. Maybe you'll finish it to:

- Improve its appearance
- Increase its durability by increasing its resistance to wear, corrosion, heat, or other elements
- Clean the rough surfaces
- Change its size and shape
- Improve its electrical conductivity

These are only a few ways surface finishing can be useful. Although many of the same reasons for using a surface finish apply to both 3D-printed parts and parts created using more conventional methods, you should also recognize the specific features of your part when selecting the surface finishing processes to be used. The manufacturing process that you use will have an effect on these features, which may include:

- The material that your part is made of
- The shape of the part

- Its thickness
- Its weight

Its expected function and the environment will be exposed to when used.

In fact, the surface finishing of 3D-printed parts should be part of your design process long before you have a finished product, as this step can be an integral part of the product's functionality. Below are some of the surface finishing processes that you can consider when making parts using 3D-printing technology.

7.2.1 PLATING

Plating means coating a metal surface with a plastic or metal substratum by subjecting the surfaces either to electrical current or to a chemical solution. The metal forms used for the plating differ. On the first layer of plastic plating you will most likely use nickel or copper, or even gold and silver. After that, you can apply almost any metal, including:

- Platinum
- Chromium
- Tin
- Palladium
- Rhodium

The best metal for your project depends on which features you want to upgrade or add to your printed component. Plating, whatever method or metal you use, has a number of advantages when used with 3D-printed parts. Many of the components produced by this technique are made of plastic, which has its advantages but leaves the part susceptible to damage due to impact, wear, and other external factors. A metal coating can enhance its strength and durability by providing a protective exterior layer. A metal part can be used to improve resistance to corrosion, oxidation, wear, and other factors, as well as to increase strength. This finishing technique will enhance a printed item's appearance too. The item may have an uneven color or a slightly uneven surface coming out of the printer. Covering the ground with a smooth, polished layer of metal offers a touch of beauty and sophistication. New properties such as electrical conductivity and thermal transfer properties can also be applied to your component by a metal sheet.

7.2.2 SANDING

Sanding is a surface finishing technique that is well known. It involves using rough material such as sandpaper to smooth out and remove small imperfections from the surface.

When you remove any 3D-printed items from the printer, one can see the thin lines where each new layer begins. It does not matter for certain industrial products, but for consumer goods, prototypes, and other things on display, when the product is to be as esthetically appealing as possible. Sanding eliminates imperfections and gives a surface that is smooth.

Sanding comes in handy when planning to apply any kind of coat to the surface because it's going to have to be smooth so that the coating comes out evenly. While sanding is effective, one downside is that it can be time consuming, especially if done by hand. Some spots, particularly small holes and undercuts, can also be difficult to reach.

7.2.3 BEAD BLASTING

Bead blasting is more effective for those hard-to-reach areas. In this process, you use a spray gun to fire finely ground thermoplastics at the surface, blow away imperfections, and smooth out the surface in a manner similar to sanding. Bead blasting, however, can be done in a much shorter time and can reach the inside of the channels and other tricky spots.

As with sanding, bead blasting is useful for enhancing the product's esthetics. It will result in a smooth, matt surface. This is also an easy way to smooth out the substance before applying the coating so that it adheres correctly. If you are using bead blasting, be sure to start at low pressure and slowly increase it if you need to stay too long in any place. Pressure that's too high or that stays on one spot for too long can blow away too much of your part and create small divots.

7.2.4 SHOT PEENING

Shot peening is a method similar to bead blasting, but it is used with a specific target in mind. Other than removing imperfections or unnecessary defects from the surface of the product, this technique is primarily used to improve the strength and durability of the part.

Pressurized air fires tiny metal or plastic beads on the surface at high speeds, similar to bead blowing. These particles create small dimples on the surface of the object under which the stress of compression is formed, and as the beads bomb the component, the dimples begin to overlap.

The compressive stress, the type of stress that decreases the length of the object, replaces the tensile stress in the object, which increases the length of the object. The stress of compression makes the surface stronger and helps to resist fatigue, wear, cracking, and cavity erosion.

7.2.5 HEAT TREATMENTS

3D-printed objects, whether metal or plastic, may also undergo one of several types of heat treatment as part of the finishing process. Like shot peening, heat treatments improve the material by reducing the stress strain. It compensates for the high and low stress areas of the item, helping to avoid failure. You may also use heat treatment to increase density or help shape the item to the final desired shape.

Following more normal heat treatments, which reduce tension and increase energy, some artifacts may undergo hot isostatic pressing. Also called HIP, this heat treatment eliminates pores and repairs any defects before the part reaches 100% theoretical density. This step is critical if the part is exposed to some harsh conditions or if fatigue failure is likely to have serious consequences. This applies to components which will be used in sectors such as:

- Aerospace
- Marines
- Medical
- Generation of power

Initial heat treatment normally takes place in a vacuum furnace, while HIP generally occurs in a pressure vessel. However, some equipment can perform both functions and use rapid cooling.

7.2.6 VIBRATORY SYSTEMS

Vibration systems can be used to process and polish multiple 3D-printed objects at once. To use this process, position the items to be polished in a unit that contains a ground-up material that is smoother than the 3D-printed component. This material is often ceramic, plastic, corn cob, or synthetic medium. The vibratory unit is usually tubular or bowl-shaped and may come in a variety of sizes. The unit pulsates until the medium polishes the components repeatedly.

This technique is most commonly used for metal components because the polishing medium must be softer than the component, but it can also be used to polish different types of heavy plastics. The biggest benefit is that you can finish several items at once with minimum manual effort.

One downside is that this method can damage uneven edges and corners of the piece. Because of this, it is mostly applied to round products or those with rounded edges. This method is not as reliable as some others, because it finishes the whole surface at once.

7.2.7 TUMBLING

Tumbling is similar to vibration, except the components and the polishing medium rotate around in the drum instead of vibrating. It is a gentler movement that makes tumbling perfect for the more delicate sections and those with the finest details. The unit used is sometimes referred to as a centrifugal barrel device. The same polishing materials are used for tumbling as vibratory systems—ceramic, rubber, corn cob, and synthetic materials. You can also finish several sections at once, as with vibration. Nevertheless, you need to be vigilant when combining various types of media, as certain combinations cause inconsistency and result in uneven finishing or harm to pieces.

7.2.8   VAPOR SMOOTHING

Vapor smoothing is another method of removing the surface of printed piece, but results in a glossy finish rather than a matte one. This sheen can be removed, however, if desired, with bead blasting or sanding. The vapor smoothing process uses a solvent to melt the surface of your component. After putting the product in a vapor chamber where it is exposed to a solvent, you immediately place it in a cooling chamber to avoid the liquefaction. This cooling ensures that only the surface is melted and the form of the object is preserved.

This method often fills any holes in the exterior of the product and seals the surface, making it especially useful for products intended to hold liquids or gases. Vapor smoothing cannot be used on any form of material, as it can cause harmful chemical reactions in certain materials, including polycarbonate, polyphenylsulfone (PPSF), ULTEM 1010, and ULTEM 9085. Of course, you will also need a chamber that fits your object, which can be restrictive when printing large components in particular.

7.2.9   SOLVENT DIPPING

The alternative to vapor smoothing is solvent dipping, which, as the name suggests, calls for the part to be dipped into a solvent rather than exposed to a vaporized solvent. This method may be useful if the component is bigger than the size of the vapor chamber.

The findings are very similar to those of vapor smoothing, but it is more difficult to preserve dimensional precision since the solvent works more rapidly and aggressively.

7.2.10   EPOXY COATING

The use of a solvent is not the only choice during the 3D-printing process for sealing a part's surface. You can also apply an epoxy coating to the surface of your item to create an airtight seal around it which will also enhance high temperature resistance to various chemicals. This approach is ideal for components that face harsh operating conditions. The epoxy coating is typically applied manually. Although this lowers costs as you do not have to buy costly equipment, it increases the time and energy required to install it and makes it more suitable for limited manufacturing runs, small parts, or products that need just a portion of their sealed surface. You may also not have access to certain areas, such as internal channels and undercuts. It may also not be ideal for parts requiring very precise dimensions, as the epoxy coating adds a small amount of thickness to the part.

7.2.11   EPOXY INFILTRATION

Nonetheless, there is another form of epoxy application available that addresses many of the drawbacks of manual epoxy application. In the case of epoxy infiltration, you plunge the product into an epoxy resin and use a vacuum chamber to draw the resin into the part so it can fill any pores. The seal formed by epoxy infiltration is airtight and watertight, as well as high temperature and resistant to many chemicals. The method takes about three hours to complete and requires less manual labor, which means that it is more efficient for large parts and large amounts of parts than to apply epoxy by hand. You can also avoid significant changes in dimensions if the process is carried out precisely. The key drawback of epoxy infiltration compared to the application of hand coating is its higher cost. You need a vacuum chamber, an oven to pre-heat and cure the resin, and the cost of the epoxy resin is also higher.

7.2.12 PAINTING

Another common finishing technique is painting, which improves esthetic appeal and also has certain sealing properties, but it will not be airtight and resistant to high temperatures and chemicals like more durable sealing methods. Although the methods and effects of applying 3D-printed paint differ greatly depending on its intended application, the key motive behind painting a product is usually an esthetic one.

Often, before you add paint, the object is likely to have undergone various other processes to smooth, seal, or otherwise finish it. Once you paint, you may want to add a primer and fill any pores with body putty, which must be sanded down to make the surface smooth again. Then you could put on another coat of primer before applying one or more layers of paint and, finally, a clear coat to protect the paint.

7.3 APPLICATIONS IN DESIGN

7.3.1 CAD MODEL VERIFICATION

The initial objective of the designers is the need for the physical part to confirm the design of the CAD system. The parts or products to be designed shall be verified whether or not the esthetic functions of the printed parts are fulfilled.

7.3.2 VISUALIZING OBJECTS

Models built on CAD systems need to be easily visualized, generated, and sold between designers and other departments. Simple visualization of objects allows all those people in every conversation to refer to the same thing.

7.3.3 PROOF OF CONCEPT

Proof of concept applies to adaptation, to specific details of the material environment, to esthetic aspects, and to specific details of the design concerning the practical performance of the material.

7.4 APPLICATIONS IN ENGINEERING, ANALYSIS, AND PLANNING

7.4.1 SCALING

Rapid prototyping technologies allow simple scaling down (or up) the size of the model by scaling the original CAD model. In the case of designs with different holding capacities, the designer can simply scale the CAD model to the desired capacity and display the renderings on the CAD program.

7.4.2 FORM AND FIT

Sizes, volumes, and forms must be considered from an asthetic and functional point of view. How a part fits into a design and its environment are important aspects that need to be addressed. The model will be used to determine how it meets both the visual and functional requirements. Form and fit models are used in the automotive, aerospace, consumer electronics, and appliances industries.

7.4.3 FLOW ANALYSIS

The designs of products that influence by air or fluid flow cannot be easily changed if they are manufactured by conventional manufacturing methods. The original 3D concept data can be stored in a virtual model and any adjustment in object data based on different tests can be made with virtual help. Flow analyses are necessary for products manufactured in the aerospace, automotive, biomedical, and shipbuilding industries.

7.4.4 PRE-PRODUCTION PARTS

In cases where mass-production is implemented after the design of the prototype has been tested and validated, pilot-production runs of ten or more parts are typical. Pilot-production parts are used to validate the design and specifications of the tooling. Many of the rapid prototyping methods are capable of producing pilot parts quickly, thus helping to shorten the process development time and thus accelerate the overall time-to-market process.

7.5 APPLICATIONS IN MANUFACTURING AND TOOLING

7.5.1 DIRECT SOFT TOOLING

This is where the mold tool is created directly by fast prototyping systems. These methods may be used for liquid metal sand casting in which the mold is broken by a single casting.

7.5.2 INDIRECT SOFT TOOLING

In this fast tooling process, a master pattern is first created using rapid prototyping. Using the master design, the molding tooling can be made from a range of materials such as silicone rubber, epoxy resin, low melting point metals, and ceramics.

7.5.3 DIRECT HARD TOOLING

Hard tools produced by rapid prototyping systems have been a major research area in recent years. The main advantage of hard tooling produced by rapid prototyping methods is a fast turnaround time to create highly complex molding tools for high-volume production. The rapid response to changes in standardized designs can be almost instant.

7.5.4 INDIRECT HARD TOOLING

Indirect hard tooling approaches use fast prototyping aid in a variety of ways. Many of these processes remain largely similar in nature, with the exception of small differences in the binding system formulations or the type of system used. Processes include the rapid solidification process (RSP). Indirect methods for manufacturing hard tools for plastic injection molding include using liquid metal casting or steel powders in a binder system.

7.6 AEROSPACE INDUSTRY

3D-printing technology provides unparalleled versatility in product design and development. 3D-printing technology in the aerospace industry has the ability to manufacture lightweight components, enhanced and complex geometries which can reduce energy requirements and resources. At the same time, it can lead to fuel savings using 3D-printing technology, as it will reduce the material used to manufacture aerospace components. In addition, 3D-printing technology has been commonly used in the manufacture of spare parts for some aerospace products, such as engines. The engine part is easily damaged, requiring regular replacement. 3D-printing technology is also a good option for the production of these spare parts. In the aerospace industry, nickel-based alloys are preferred due to their tensile properties, oxidation/corrosion resistance, and damage tolerance (Figure 7.1).

The aerospace and defense (A&D) industry was one of the first to embrace 3D printing, with the first application of technology dating back to 1989. Now, three decades later, A&D represents 12% of the $7 billion additive manufacturing market and contributes heavily to ongoing research efforts within the industry.

The advancement of AM within A&D is largely driven by key industry players, including GE, Airbus, Boeing, Safran, and GKN. Such businesses and others have described the value proposition that 3D printing brings to:

- Functional prototypes
- Tooling
- Lightweight components

As we can see, aerospace 3D printing is not limited to prototypes. Actual, functional components are often printed in 3D and used in aviation. Examples of components that can be produced with 3D printing include air ducts (SLS), wall panels (FDM), and components of structural metal (DMLS, EBM, DED).

FIGURE 7.1   3D-printed aerospace parts.

7.6.1   The Benefits of 3D Printing for Aerospace and Defense

7.6.1.1   Low-Volume Production

3D printing is suitable for industries such as aerospace and defense, where highly complex parts are manufactured in small quantities. Using technology, complex geometries can be created without having to invest in expensive tools. It gives aerospace OEMs and manufacturers a cost-effective way to manufacture small batches of parts in a cost-effective manner.

7.6.1.2   Weight Reduction

Including aerodynamics and engine efficiency, weight is one of the most important factors to be considered when it comes to designing an aircraft. Reducing aircraft weight will reduce carbon dioxide emissions, fuel consumption, and payload significantly. That is where 3D printing comes in: Technology is the perfect method for producing lightweight parts, resulting in significant fuel economy. When paired with design optimization tools such as generative design software, the component's potential to increase complexity is nearly infinite.

7.6.1.3   Material Efficiency

Since the 3D printing process works by creating parts layer by layer, most of the material is used only where it is required. As a consequence, less waste is produced than traditional subtractive methods.

The selection of available aerospace and defense 3D printable materials varies from engineering thermoplastics (e.g. ULTEM 9085, ULTEM 1010, PAEK, reinforced nylon) to metal powders (high-performance alloys, titanium, aluminum, and stainless steel). The variety of 3D printable materials available is constantly expanding, allowing advanced aerospace applications.

7.6.1.4 Consolidation of the Part

One of the main benefits of 3D printing is the aspect of convergence: The ability to combine several parts into one piece. Reducing the amount of parts available would minimize the assembly and repair period considerably by reducing the time needed for assembly.

7.6.1.5 Maintenance and Repair

The average lifespan of an aircraft can range from 20 to 30 years, making maintenance, repair, and overhaul (MRO) an important function in the industry. Metal 3D-printing techniques such as direct energy deposition are widely used to repair aerospace and military hardware. Turbine blades and other high-end devices may also be restored and repaired by applying material to the worn-out surface.

7.7 AUTOMOTIVE INDUSTRY

Nowadays, 3D-printing technology has drastically changed our industry to design, develop, and manufacture new things. In the automotive industry, 3D-printing technology has created phenomena that carry new shine, allowing for lighter and more complex structures in a fast time. For example, the first 3D-printed electric car was printed by Local Motors in 2014. Local Motors has also expanded the wide range of applications of 3D-printing technology to the maker of a 3D-printed bus called OLLI. OLLI is a driverless, electronic, recyclable, and extremely smart 3D-printed vehicle. In addition, Ford is a leader in the use of 3D-printing technology and also applies 3D-printing technology to the production of prototypes and engine parts. In addition, BMW uses 3D-printing technology to develop hand tools for vehicle research and assembly. Meanwhile, in 2017, Audi cooperated with SLM Solution Group AG in the development of spare parts and prototypes. As a result, the use of 3D-printing technology in the automotive industry enables the company to try out different alternatives and to emphasize right at the stage of improvement, creating an ideal and efficient automotive design. At the same time, 3D-printing technology can reduce material waste and consumption. In addition, 3D-printing technology can reduce costs and time, making it possible to test new designs in a very fast time.

The automotive industry is a growing user of additive manufacturing: In 2018 alone, the automotive 3D-printing market was estimated to be worth $1.4 billion. This figure is only set to increase, as the market is projected to reach $5 billion by 2023, according to one study. Across fields such as motorsports and performance racing, techniques such as generative design and topology optimization are slowly changing traditional approaches to parts design. While prototyping continues to be the main application of 3D printing in the automotive industry, companies are increasingly finding other applications, such as tooling. In addition, many automotive companies are beginning to find creative end-use applications for 3D printing, which represent an exciting trend for the industry (Figure 7.2).

FIGURE 7.2   3D-printed automotive engine parts.

7.7.1 THE BENEFITS OF 3D PRINTING FOR AUTOMOTIVE

7.7.1.1 Faster Product Development

Prototyping has become a vital part of the product development process, providing a way to check and verify the parts before they are made. 3D printing provides a simple and cost-effective approach to the design and manufacture of components. When the need for tools is reduced, product teams will greatly shorten product development cycles.

7.7.1.2 Greater Design Flexibility

The ability to produce designs quickly gives designers more flexibility when testing multiple design options. 3D printing allows designers to make quick design changes and modifications within a fraction of the time.

7.7.1.3 Customization

3D printing offers a cost-effective and flexible way for automakers to produce customized parts. In the luxury and motorsport segment of the market, businesses are now using the technology to manufacture custom parts for both the interior and exterior parts of a car.

7.7.1.4 Create Complex Geometries

With the majority of car components requiring complex geometries such as internal channels (for conformal cooling), thin walls, and fine mesh, AM makes it possible to produce highly complex parts that are still lightweight and durable.

7.8 JEWELRY INDUSTRY

An unexpected use of 3D printing is in the apparel industry. 3D-printed jewelry has become a popular niche for those looking for a unique look. With the advent of 3D printers, jewelers can experiment with designs and not limited to using conventional jewelry-making methods. In addition, 3D printers make it cheaper to manufacture individual, unique pieces of jewelry or customize pieces for customers. Jewelry manufacturing is one industry that embraces 3D printing. Most jewelers now use technology to challenge the way things have been done for decades. Usually 3D printing is used to make jewelry using two methods: investment casting and direct printing.

7.8.1 INVESTMENT CASTING

The investment casting process is one of the most common methods of producing jewelry via 3D printing. Parts are created by investment casting through an eight-step process:

1. Pattern formation: This was traditionally done by pouring a special cast of wax into a metal mold. The 3D printing also allows direct printing of the pattern from wax or castable resin (Figure 7.3).
2. Mold assembly: The molded or printed pattern is then assembled on a "casting tree." This allows multiple parts to be cast at the same time. Some 3D-printing methods disrupt this step by printing part patterns and a tree in one step (Figure 7.4).
3. Shell building: Upon completion of the pattern assembly, the entire assembly is immersed in slurry several times. The slurry coating is then left to dry and solidify, forming an outer ceramic layer over the pattern (Figure 7.5).
4. Burnout: The structure is then put inside the furnace and the original wax/resin structure is melted/burned resulting in a hollow negative mold (cavity) (Figure 7.6).
5. Pouring: When all the original pattern material has been removed from the negative ceramic, the final casting material is poured into molds and left to cool and solidify. Parts are often cast in brass and electroplated in precious metals during the finishing stage (Figure 7.7).

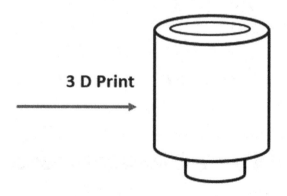

3 D Print

FIGURE 7.3 Pattern formation process.

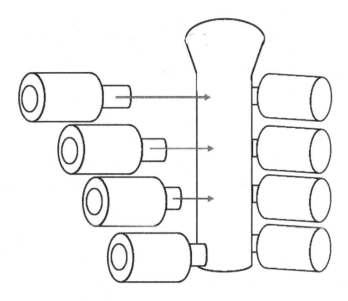

FIGURE 7.4 Mold assembly process.

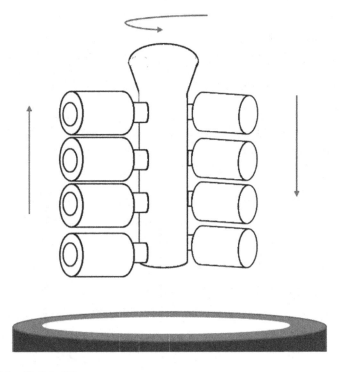

FIGURE 7.5 Shell building process.

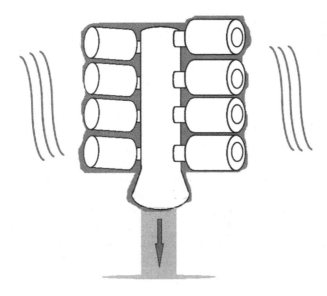

FIGURE 7.6 Burnout process.

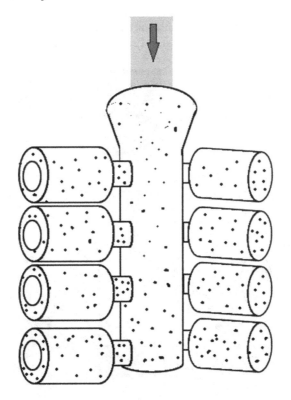

FIGURE 7.7 Pouring process.

6. Knock off: The outer ceramic mold must be removed. This is usually done by vibrating the mold to knock off the outer shell (Figure 7.8).
7. Cut off: Once the ceramic shell has been fully removed, the individual cast objects are cut off from the mold tree (Figure 7.9).
8. Finishing: The cast parts then go through traditional jeweler finishing techniques (Figure 7.10).

There are many criteria for 3D-printing technology for the efficient development of jewelry molds for investment casting. These are:

- Engineering must be capable of manufacturing products with a very high degree of detail and minute, intricate features.
- At the burnout/melt stage, the material used to print the pattern must be completely removed. The remains of the original pattern material have a negative effect on the consistency of the final cast product. Due to this, strict burnout procedures are in place for most 3D-printed castable resins (Figure 7.11).

7.8.2 DIRECT PRINTING

A much less common method of manufacturing jewelry by 3D printing is the direct printing of parts made of metal powder. Parts may be printed by means of gold,

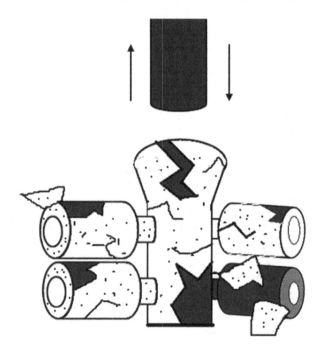

FIGURE 7.8   Knock off process.

FIGURE 7.9 Cut off process.

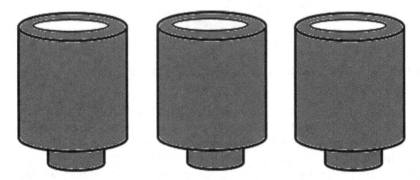

FIGURE 7.10 Finishing process.

silver, or platinum alloys and then require a large amount of post-processed finishing. Direct jewelry printing is generally more expensive than investment casting, even for single pieces, and requires a very high level of precious powder management. DMLS/SLM, direct metal laser sintering (DMLS), or selective laser melting

FIGURE 7.11   A cast jewelry tree before removal of the parts from the tree.

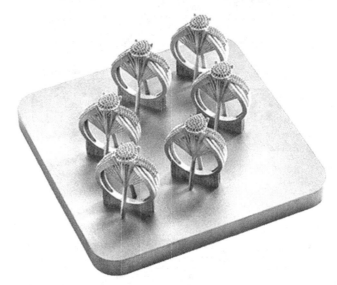

FIGURE 7.12   Jewelry produced via metal 3D printing still attached to the build plate and showing support material.

(SLM) are powder bed melting techniques used in the production of metal parts. For the accurate production of DMLS/SLM pieces, a large amount of support must be provided during printing. High temperatures result in high levels of stress, meaning that parts are often susceptible to warping or deformation. This leads to the need for significant post-processing to remove the support and finish the surface where it was attached (Figure 7.12).

7.9   COIN INDUSTRY

The value of 3D printing is greatest when used to produce the coveted "impossible objects." Direct precious metal 3D printing has best demonstrated this in the jewelry sector, where the process has opened up unique design and weight-loss capabilities. So what happens when you introduce this level of innovation to a completely different traditional industry like coin making?

Cooksongold, part of the Heimerle + Meule Group, has a long history of producing precious metal products for the jewelry and watchmaking industries. In 2014, in collaboration with EOS, the company introduced the Precious M 080, an advanced manufacturing technology that allows the user to create complex jewelry and watch components in the Advanced Precious Metal Powder range: 18k gold, 950 Platinum, and 925 Silver.

In its most recent project, technology has been used to tackle another type of industrial production—the mint. Cooksongold already supplies blank coins to a number of mints around the world, which then hit the coins with their own images. With that, the supplier of precious metals decided to set itself the challenge of printing the world's first truly 3D image directly onto the face of an existing blank coin using a different alloy.

To accomplish this, the building foundation was first milled to house the current 18k yellow gold coin blank. Using a CAD image of the "crown" design, the image was 3D printed in 20-micron layers using 500 g of 18k white powder directly on the coin. The 3D-printing process of precious metal melted the powder directly to the surface of the coin, ensuring a strong bond between the metals. Using a special manufacturing method, it was possible for the coin to have undercuts and display a true 3D image that is simply not possible using conventional stamping techniques (Figure 7.13).

FIGURE 7.13   3D-printed coin.

7.10 TABLEWARE INDUSTRY

As 3D-printing technology continues to improve, consumers suddenly find the ability to manufacture their own products. While 3D desktop printing has its limitations, there are a lot of useful objects that can be created right at home.

There have been 3D ceramic printers in the past, but most of them have been printed in very poor materials, with information that leaves a lot to be desired. Such printers have never been able to print useful, reliable tableware, such as teacups, bowls, etc.

The world's first 3D ceramic tableware printer was created by Bristol University students. The printer from the University of West England currently prints a porcelain content much superior to what has been seen with previous printers. The machine will be very attractive to artists, designers, and manufacturers of tableware and other ceramic products. The entire process of printing, glazing, and firing each piece may still take a few days, but those in the ceramics industry are used to such waiting times (Figure 7.14).

7.11 GEOGRAPHIC INFORMATION SYSTEM (GIS) APPLICATIONS

The geographic information system (GIS) is a system designed to collect, store, process, interpret, handle, and present spatial or geographical data. GIS applications are tools that allow users to create interactive queries (user-generated searches), analyze spatial data, edit data in maps, and present the results of all these operations. GIS (more commonly GIScience) sometimes refers to geographic information science (GIScience), the science behind geographic concepts, applications, and systems. Since the mid-1980s, geographic information systems have become a valuable tool used to support a variety of urban and regional planning functions.

Similar to 2D maps, 3D GIS maps represent objects in more detail by adding another dimension (z). The 3D technology of GIS maps is an explanatory representation of the size of real-world objects. 3D models assist in the presentation of surveys in a large number of different domains. For example, 3D maps can show the height of a hotel or a mountain, and not just its location. The 3D tools must be used along with the 2D GIS and then configured in a 3D setting.

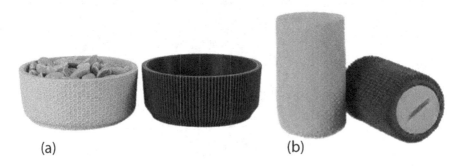

(a)   (b)

FIGURE 7.14   (a) Bowl set; (b) Salt and pepper shaker.

There was a time when more than one program was required to view objects on the streets and in different parts of the city. The modern geographic information system (GIS) has changed the dynamics of Geographic and Earth Sciences. With the advent of a digital medium, the modern GIS interface allows its users not only to envisage and analyze, but also to handle spatial facts and figures.

Over the years, GIS has had a significant impact in creating mapping as a key tool to solve problems. Conventionally, GIS information was based on a two-dimensional recording, which apparently limited its use in most applications. The incorporation of 3D technology in GIS customizes the entire experience, making it more personal and enabling detailed visualization (Figure 7.15).

Let us look at some of the applications of 3D GIS:

i) City planning: Today, most cities face a shortage of basic amenities such as water, electricity, and living space. The issues can be attributed to an incorrect allocation of resources. Incorporating 3D technology into GIS can help government agencies, architects, and engineers plan, evaluate, and analyze how certain changes in the city will look and how they can meet the needs of future generations. A typical 3D model will consist of construction details, satellite imagery, and traffic data that urban planners would use to efficiently identify potential approaches and resolve emergency situations efficiently.

ii) Building information modeling: Building information modeling (BIM) is an important technology that depicts the settings of the real-world environment. The integration of BIM and GIS offers the requisite know-how

FIGURE 7.15   Geographic information system (GIS).

to build a robust model. The combination of 3D GIS and BIM can help to produce error-free building management plans that would eventually allow for a more detailed analysis of the data.

iii) Coastal modeling and analysis: Coastal areas are important because they connect a country to the rest of the world for trade. Globally, coastal areas face major threats and construction problems. It is important for planners to recognize what all factors affect the construction and conservation of ports, fisheries, and mining operations. Efficient and effective 3D GIS resource planning can provide some level of understanding in the economic and environmental movements along the coast.

iv) Disaster response: 3D GIS can help people and societies better deal with natural disasters. In the event of a disaster, accurate mapping will give a broad understanding to disaster response teams by making them aware of the area in which they would be working. It will also require the collection of details such as:

- Precise coordinates of the conflict point and the fastest way to get there
- Description of the environment, including the type of terrain
- An alternate path for safe evacuation

7.12 ARTS AND ARCHITECTURE

Trained architects, engineers, and construction (AEC) recognize the essential value of reliable and practical scale models. It helps them and their clients to visualize ideas realistically and vividly. Modern methods of designing and constructing scale-models are time-consuming, expensive, and rely heavily on a handful of professional craftsmen.

3D printing is intended to revolutionize the way architects approach architecture and innovate. 3D printing for architects enables them to easily create complex, accurate, and durable scale models quickly and cost-effectively. Magnificent 3D-printed architectural scale models will help architects attract their clients by creating more opportunities. This can be done in-house, in a matter of clicks (Figure 7.16).

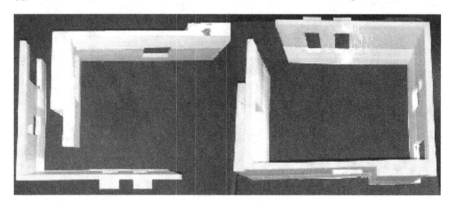

FIGURE 7.16 3D-printed architect design.

7.12.1   Benefits of 3D Printing for Architects

Many leading model manufacturers and architectural firms have already reaped the benefits of 3D printing in architecture. 3D-printed architectural scale models accurately reflect the final appearance of the design, rendering the design visible, leaving a lasting visual impression. The following advantages are offered by 3D printing for architects:

- **Save time and money**: One of the key advantages of 3D printing for architects is time-saving and cost-effectiveness. In contrast to traditional methods, 3D-printed architectural scale models can be developed in a matter of hours. Conventional methods take several days, several man-hours, and professional craftsmen, adding to the expense.
- **Seamless integration**: Many architectural companies now have in-house design teams that use CAD applications. 3D printers can easily interact with these applications in order to accurately make scale models without the presence of human error, thereby seamlessly incorporating them into the design process.
- **Added design possibilities**: 3D printers allow architects to design freely without worrying about human error in the final production. 3D-printed architectural scale models are impeccably accurate. This freedom empowers architects to push the boundaries of design while allowing multiple copies to be rendered faster than ever before.
- **Better perspective**: No number of drawings, blueprints, or digital 3D models can replicate the "real-life" viewpoint provided by 3D-printed architectural scale models. Architects can identify, test, and evaluate the scale model for design flaws by taking pre-construction corrective action. These visually attractive scale models can also be used in promotions and presentations to customers.

7.13   CONSTRUCTION

3D printing offers a variety of technologies that use 3D printing as the main means of constructing buildings or building components. 3D-printing technologies used on a construction scale include extrusion (concrete/cement, wax, foam, and polymers), powder bonding (polymer bonding, reactive bonding, sintering), and additive welding. 3D printing in construction has a wide range of applications in the residential, commercial, industrial, and public sectors. The advantages of these technologies include allowing for more complexity and accuracy, faster construction, lower labor costs, greater functional integration, and less waste.

The first fully finished 3D-printed residential building was built in Yaroslavl, Russia, in 2017. 600 elements of the walls were printed in the shop and assembled on site, followed by the completion of the roof structure and interior decoration for a total area of 2985 square meters (3213 sq. ft.). This project is the first in the world whereby the entire technological cycle had passed on building requirements, from

FIGURE 7.17 Largest 3D-printed building.

design, building permits, registration, to the integration of all engineering systems. The building wasn't built just for presentation; it's a real building, with normal families living there today.

Concrete 3D printing has been in progress since the 1990s as a faster and cheaper way to build buildings and other structures. Large 3D printers designed specifically for printing concrete can pour foundations and build walls on site. They can also be used for printing modular concrete parts, which are later installed on the job site.

In 2016, the first pedestrian bridge was printed in 3D in Alcobendas, Madrid, Spain. It was printed in micro-reinforced concrete with a length of 12 meters (39 ft.) and a width of 1.75 meters (5.7 ft.). The bridge illustrates the complexities of nature and has been designed for both parametric (using a set of rules, values, and relationships that guide the design) and computational design, allowing for an optimal distribution of materials while maximizing structural performance.

The first large-scale application of 3D-printing technology in the field of civil engineering in a public space was a landmark in the international construction industry.

3D printing is used to create architectural scale models, allowing a quicker turnaround of the scale model and increasing the overall speed and complexity of the artifacts produced (Figure 7.17).

7.14 FASHION AND TEXTILES

The growth in fast fashion already has disrupted the seasonality of the fashion industry, and 3D printing has the power to further accelerate production. It also allows consumers to get involved in designing of the clothes they wear. Because 3D printing works well with hard materials, it has been introduced to the fashion industry with jewelry, shoes, and ornamentation. Brands like Adidas, Reebok, and New Balance have all launched 3D-printing initiatives.

On the jewelry front, Lockheed Martin filed a patent application for a 3D printer that produced synthetic diamonds in 2016. Another company, Sandvik, prints diamonds using composite materials. While the applications for composite diamonds are manufacturing-focused (e.g. drills), the technology could eventually spread to consumer diamonds. Many fashion houses, enabled by 3D-technology companies such as CLO, are now using 3D scanning and 3D printing to create custom products. Designer Iris Van Herpen unveiled a collection of 3D-printed clothes during the Paris Fashion Week 2018 show.

Despite the potential of 3D printing in the fashion industry, there are limitations to what it can do with soft materials and non-geometric shapes. Some envisage a hybrid future in which 3D printing works in tandem with traditional methods to leverage the best of both worlds (Figure 7.18).

FIGURE 7.18   3D-printed textile.

7.15 WEAPONS

In 2012, Defense Distributed, a US-based group, announced plans to "[design] a working plastic gun that could be downloaded and reproduced by anyone with a 3D printer." Defense Distributed also designed a 3D printable AR-15 type rifle lower receiver (capable of more than 650 rounds in duration) and a 30-round M16 magazine. The AR-15 has multiple receivers (both the upper and lower receiver), but the legally controlled part is the one that is serialized (the lower part, in the case of the AR-15). Soon after Defense Distributed succeeded in designing the first working blueprint for the production of a plastic 3D printer gun in May 2013, the US Department of State requested that the instructions be removed from their website. After Defense Distributed released its plans, questions were raised about the potential impact of 3D printing and widespread consumer CNC machining on gun control effectiveness.

In 2014, a man from Japan became the first person in the world to be jailed for making 3D-printed weapons. Yoshitomo Imura posted online videos and gun blueprints and was sentenced to two years in prison. In his home, the police found at least two weapons capable of shooting bullets (Figure 7.19).

FIGURE 7.19   3D-printed weapons.

7.16 MUSICAL INSTRUMENTS

The shape and material of the instrument influences the sound it makes. Although 3D printing has yet to see substantial progress in the music industry, it may allow for new possibilities for instrument design, composition, and sound.

FIGURE 7.20 3D-printed guitar.

Instrument makers have started experimenting with 3D printing, revealing inventions such as a titanium violin and a range of custom string instruments.

Printed tools will become more popular as 3D-printing technology evolves to work with more materials outside of plastic and metal (Figure 7.20).

7.17 FOOD

The 3D printer is an exciting tool that produces three-dimensional artifacts. The printer creates an item by placing the printing medium in layers. Instead of using ink as a medium, many consumer level 3D printers use melted plastic that solidifies almost immediately after it is released from the printing nozzle. However, other printing media are available, including a relatively new one—powdered or liquid food material. Sugar, liquid chocolate, and puréed food have all been used to create new food items with interesting and complex shapes and patterns. For certain situations, the use of a 3D printer to manufacture a food item is simpler than the creation of a food item by hand.

In the near future, 3D food printers may have additional benefits. NASA has partnered with a Texas company to develop a more capable printer type. The printer would be able to mix powdered material with liquid to create a wide range of foods. The aim of NASA is to increase the quality, stability, and protection of food provided to astronauts while in space. It will be especially important during deep space

FIGURE 7.21   3D-printed food.

missions. It has been proposed that the new printer could also reduce world hunger. Experimentally, another form of food printer was used to produce meat (Figure 7.21).

7.18   MOVIES

3D-printing technology is changing almost every area of our lives. It is used in medicine, in the automotive sector, in architecture, and now also in the film industry. 3D printing enables the creation of unique props in a fast and extremely accurate manner. This is crucial when you expect realism on screen, but also science-fiction. 3D printing allows directors to bring their wildest visions to life.

The first time you watch a movie, you probably just want to get swept away and get caught up in an adventure. It's only later when you let the photos tumble in your mind that you start to think about it on other levels.

You might think of the weapons, the various objects that are integral to the story, the animated figures, the cool vehicles, or the outrageous costumes. You might wonder how they made them or how they came to look so real.

For more than a decade, 3D printing (or additive manufacturing) has been used by film industry craftsmen. In the last few years, however, the industry has rapidly expanded and become more mainstream. 3D printing has continued to offer better resolution, faster production times, and a variety of available materials—all of which are important for film making.

Film production teams often turn to professionals outside the studio to get the right skills to deliver what they need. Other times, they will acquire 3D-printing technology and use it to further streamline the process and gain control over every aspect and detail (Figure 7.22).

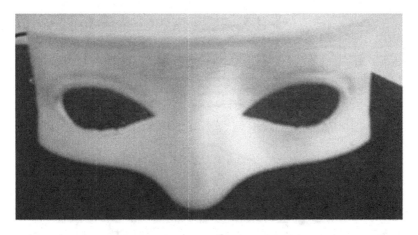

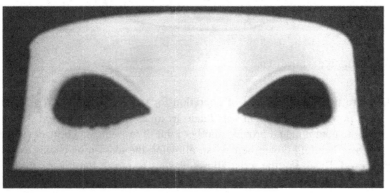

FIGURE 7.22   3D-printed props used in a movie.

7.19   DESIGN AND DEVELOPMENT OF A PROSTHETIC HAND THROUGH 3D PRINTING: CASE STUDY

Mr. Mukul Pande is a Director (IT Infrastructure) at Gaikwad-Patil Group of Institution, Nagpur (MS), India. A prosthetic robot hand is his personal project initiated with authors of this book working at Tulsiramji Gaikwad-Patil College of Engineering and Technology (TGPCET), Nagpur (MS), India. The objective of this case study was to create a freely-available, three-dimensional (3D) printable prosthetic hand. Current 3D-printed prosthetic hand designs are openly available and inexpensive to produce with a 3D printer; however, these prosthetics are also prone to failure. Tolerance issues, printing errors, and poor instructions lead to a significant number of prosthetics that cannot be assembled, do not work correctly, or break with light use.

The aim was to provide a solution to these problems through the use of equation-based scaling and proper instructions. Resizing available 3D-printed prosthetics does not always work, as holes and joints will scale with the rest of the device by the same amount, reducing functionality when larger or smaller than the initial design.

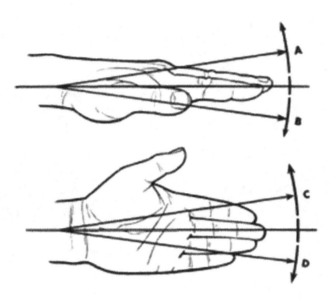

FIGURE 7.23   Types of wrist movements.

There are two main methods of operation for these prosthetics: Wrist powered and elbow powered. A decision was made to focus on wrist powered devices, as they are more common and provide another joint of movement. These devices work by the wearer bending down their wrist, allowing the tensioning cables to pull the fingers closed. Releasing the wrist allows elastics to return the fingers to a resting state. This specific prosthetic is intended for users with a moving wrist that has at least part of their palm attached to the device. The design scales using equations to scale different features at different rates, and a provided text file allows for variable editing. It is also as reliable and easy to assemble as currently available hands (Figures 7.23–7.26).

This project was initiated in June 2017 as the first 3D-printed open source prosthetic robotic hand and it has led to many projects in the area of 3D printing and its applications. This case study was done on a TEVO-Tarantula I3 Aluminum 3D printer with a 2 roll filament 8GB SD card, LCD, and extruder supports 25 filaments.

This kind of robot hand is replicable on any FDM 3D printer with a 20 × 20 × 22cm area. This idea was conceived as a development platform for the laboratory established at Tulsiramji Gaikwad-Patil College of Engineering and Technology, Nagpur, India, which is one of the most unique laboratories in central India (Figures 7.22–7.30).

This human-sized robot hand provides an appearance with five fingers and a grasping function to forearm and is entirely printed with 3D printing and has replaced the gripper of the Pick and Place 6 axis articulated industrial robot (Model: ARISTO 6XT Machine NO. 147). It provides a realistic appearance that is the same as the cosmetic prosthetic hand and a grasping function. A simple link mechanism with one linear actuator for grasping and 3D-printed parts achieve low cost, light weight, and

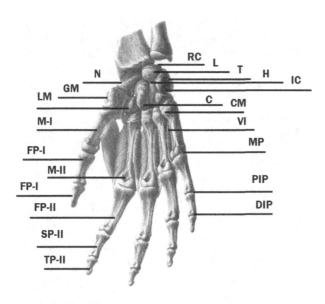

FIGURE 7.24   Bones in the hand.

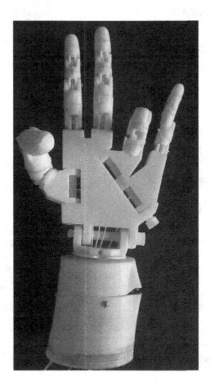

FIGURE 7.25   Case study Picture 1.1. (Courtesy: TGPCET, Nagpur, India.)

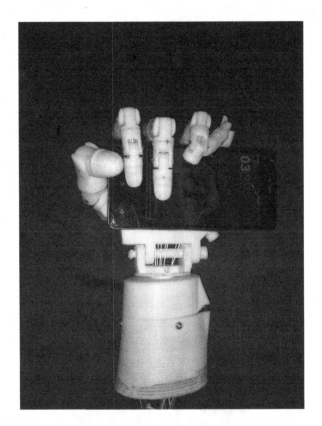

FIGURE 7.26   Case study Picture 1.2. (Courtesy: TGPCET, Nagpur, India.)

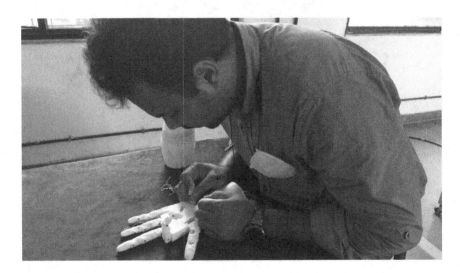

FIGURE 7.27   Case study Picture 1.3. (Courtesy: TGPCET, Nagpur, India.)

FIGURE 7.28   Case study Picture 1.4. (Courtesy: TGPCET, Nagpur, India.)

FIGURE 7.29   Case study Picture 1.5. (Courtesy: TGPCET, Nagpur, India.)

ease of maintenance. In this case each finger of the robot hand is activated through threads passed through it and synchronized with the drive motors of industrial robot and to perform the pick and place operation for the products with payload capacity of 2 kg. The sketching was performed with this 3D-printed robot synchronized with an articulated industrial robot (Figures 7.31 and 7.32).

FIGURE 7.30   Case study Picture 1.6. (Courtesy: TGPCET, Nagpur, India.)

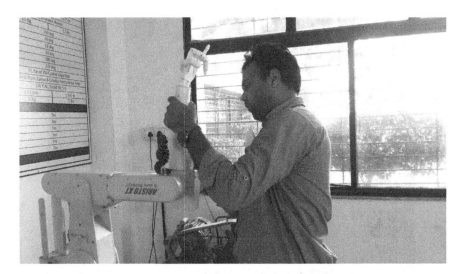

FIGURE 7.31   Case study Picture 1.7. (Courtesy: TGPCET, Nagpur, India.)

7.19.1   CONCLUSION

7.19.1.1   Prototype Assessment

Testing of the final prototype confirmed that Mr. Mukul Pande and the authors of the book, Dr. G.K. Awari and Mr. V.V. Ambade, were successful in completing the main objectives of the case study. The prototype, shown in the figures, conforms to the previously laid out design specifications. After performing a series of tests, the prosthetic was able to hold a cell phone, hold and throw a tennis ball, and it was operating with

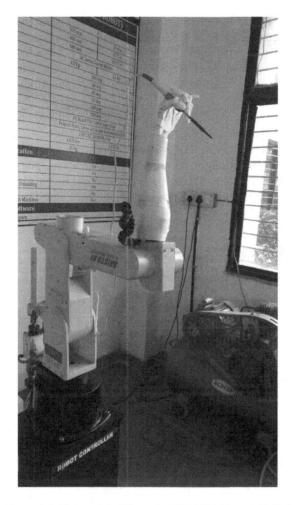

FIGURE 7.32   Case study Picture 1.8. (Courtesy: TGPCET, Nagpur, India.)

the industrial robot available in the laboratory. Through a number of design iterations, printability, ease of assembly, tolerance, and esthetics were improved.

7.20   EXERCISES

1. Discuss different materials and their material relationship.
2. Explain the finishing processes.
3. Discuss the applications of 3D printing in design.
4. Discuss the applications of 3D printing in engineering, analysis, and planning.
5. Discuss the applications of 3D printing in manufacturing and tooling.
6. Discuss the applications of 3D printing in the aerospace industry.

7. Discuss the applications of 3D printing in the automotive industry.
8. Discuss the applications of 3D printing in the jewelry industry.
9. Discuss the applications of 3D printing in the coin industry.
10. Discuss the applications of 3D printing in the tableware industry.
11. Discuss the applications of 3D printing in geographic information system (GIS) applications.
12. Discuss the applications of 3D printing in arts and architecture.
13. Discuss the applications of 3D printing in construction.
14. Discuss the applications of 3D printing in fashion and textiles.
15. Discuss the applications of 3D printing in the weapons industry.
16. Discuss the applications of 3D printing in musical instruments.
17. Discuss the applications of 3D printing in the food industry.
18. Discuss the applications of 3D printing in the movie industry.

7.21 MULTIPLE-CHOICE QUESTIONS

1. Tensile strength of polyamide (PA 6) (in MPa) ranges from
   a) 38 to 66
   b) 25 to 45
   c) 15 to 25
   d) 65 to 85
   Ans: (a)

2. Tensile strength of polyamide (PA 11) (in MPa) is
   a) 40
   b) 50
   c) 60
   d) 70
   Ans: (b)

3. Tensile strength of polyamide (PA 12) (in MPa) ranges from
   a) 21 to 31
   b) 31 to 41
   c) 41 to 48
   d) 51 to 61
   Ans: (c)

4. Tensile strength of glass bead filled polyamide (in MPa) is
   a) 30
   b) 40
   c) 50
   d) 60
   Ans: (a)

5. Tensile strength of silicon (in MPa) ranges from
   a) 1 to 2
   b) 3 to 4

c) 5 to 6
d) 6 to 9
Ans : (d)

6. Tensile strength of polypropylene (in MPa) ranges from
   a) 10 to 15
   b) 15 to 20
   c) 20 to 25
   d) 25 to 30
   Ans: (c)

7. Tensile strength of aluminum (in MPa) ranges from
   a) 310 to 400
   b) 410 to 440
   c) 450 to 480
   d) 490 to 510
   Ans: (b)

8. Tensile strength of aluminum (in MPa) ranges from
   a) 800 to 1000
   b) 1000 to 1200
   c) 1300 to 1500
   d) 1600 to 1800
   Ans: (b)

8 Additive Manufacturing Equipment

8.1 PROCESS EQUIPMENT—DESIGN AND PROCESS PARAMETERS

Each metal AM product has its own unique set of product requirements. The selection of process parameters to satisfy these product specifications is complicated by an enormous number of process parameters. There is no "Machinist Manual" for creating metal additive manufacturing (AM) products. Seven distinct AM process categories, an ever-increasing base of AM-compatible materials and material suppliers, and a wide range of process inputs, formulating the right parameters for application requires a level of expertise that is not widely available. As a matter of fact, material-unique, one-size-fits-all process parameters are typically applied.

Here is a step-by-step approach to help you select the optimum process parameters:

- Set the requirements of your product and understand the trade-offs
- Choose your AM method
- Change the configuration of your method
- Choose and qualify your feedstock
- Choose your process parameters and understand the trade-offs
- Validate the parameters of your process
- Help you identify a printing partner on a scale (if required)

8.1.1 SEVEN DISTINCT AM PROCESSES

Additive manufacturing is the method of applying 3D printing to industrial production that enables products to be manufactured without joints and with minimal post-processing. During this process, multiple materials can be used, making it easy to create new products with minimal waste and lower material costs. There are seven additive manufacturing techniques available. Each of them varies due to the materials, layering, and machine technology needed.

8.1.1.1 Powder Bed Fusion

This method of additive manufacturing uses either a laser or an electron beam to melt and fuse the powder material together to manufacture the products. Here are the variations between the two forms of powder bed melting:

- **Laser Powder Bed Fusion**: Laser powder bed fusion uses a laser to heat powdered material into 3D objects. After a layer of powder has been indexed, a new layer of powder is spread to continue the process. Ultimately laser powder bed fusion technique does not require support of other methods.

- **Electron Beam Powder Bed Fusion**: This type of powder melting bed is used to melt the particles together in specific areas. The beam can be manipulated very quickly, which speeds up the overall process by allowing multiple melt pools to occur simultaneously.

8.1.1.2 Directed Energy Deposition

In the case of directed energy deposition (DED), powder or metal wire is used with an energy source to add material or to fuse a material onto an existing part or to create a new part. Here are the three types of energy-directed deposition:

- **Laser DED**: Laser DED deposits powder on the material while the beam is melting at the same time. This process can produce much faster build-up speeds than traditional laser powder bed fusion.
- **Arc DED**: An EWI specialty that is more dynamic than any other additive manufacturing process. Arc DED is suitable for large constructions. The advantage for manufacturers is that there are existing arc-welding robots and power supply systems.
- **Electron Beam DED**: EB-DED allows incredibly quick production of large parts, which gives it an advantage over other additive manufacturing types. The process is used in heavy machinery, manufacturing, mining, and aerospace industries to produce large, low-volume components.

8.1.1.3 Binder Jetting

Binder jetting additive manufacturing uses an ink-jet printing head to print a binder on a powder that "binds" the metal particles together in a green state. The parts are then extracted from the powder bed and must undergo a de-binding and sintering process (in the oven) to make the parts completely dense and rough. Parts usually shrink by 20–25% during sintering.

8.1.1.4 Sheet Lamination

This type of additive manufacturing connects the sheets of material to form part of it. There are two types of sheet lamination additive production:

- **Ultrasonic Additive Manufacturing**: This type of sheet lamination uses ultrasonic vibrations to weld metal tapes together until it is capable of forming objects.
- **Friction Stir Welding**: Using friction stir welding improves the properties of the material when each layer is stirred. This creates diffusion and reduces the size of the grain for a secure bond.

8.1.1.5 Material Extrusion

Filament or thermoplastic material is used for the production of parts in material extrusion. The filament (or thermoplastic) is heated in this process, and then layered continuously through the nozzle to produce the final product or component. New items are available inside the plastic "rods" which are extruded with metal filler. The parts then move through the process of de-binding and sintering to produce metal bits, like binder jetting.

8.1.1.6 Material Jetting

In this additive manufacturing process, new materials are available that have a metal filler inside the plastic rods which are extruded. The parts then move through the de-binding and sintering process, like binder jetting, to create metal pieces.

8.1.1.7 Vat Photopolymerization

In contrast to other types of additive manufacturing, vat photopolymerization uses liquid resin. This photopolymer resin is applied layer by layer, and then the UV light hardens the resin to make the final part or object.

8.1.2 DESIGNING FOR 3D PRINTING

All parts produced using a 3D printer must be designed using some kind of CAD software. This type of production depends mainly on the quality of the design of the CAD and also on the precision of the printer. There are several types of CAD software available, some of which are free, others require you to purchase the software or have a membership or subscription. Selecting what kind of CAD software is perfect for you will depend on what you are designing. However, any of the free CAD software packages will do for beginners who simply want to learn CAD and create basic shapes and features.

The following points must be kept in mind when designing a part to be printed in 3D:

- The part needs to be a solid part, that is, not just a surface; it needs to have a real volume.
- The production of very small or delicate features may not be properly printed, depending on the type of 3D printer that will be used.
- Supports will be required for parts with overhanging features to be properly printed. This should be taken into account as the help needs to be removed after the model has been cleaned. This may not be an issue unless the part is very delicate, because it might break.
- Be sure to calibrate the 3D printer before using it; it is essential to ensure that the part is properly attached to the built-in plate. If it isn't, the component may be lost at some point and the whole print job may be destroyed.
- Some thought should be given to the orientation of the component, since some printers are more accurate on the X and Y axes than the Z-axis (Figures 8.1 to 8.3).

8.2 GOVERNING BONDING MECHANISM

This is aimed at describing the most critical factor governing the formation of bonds related to the ultrasonic additive manufacturing (UAM) bonding mechanism.

Creating Basic 2D Shapes:

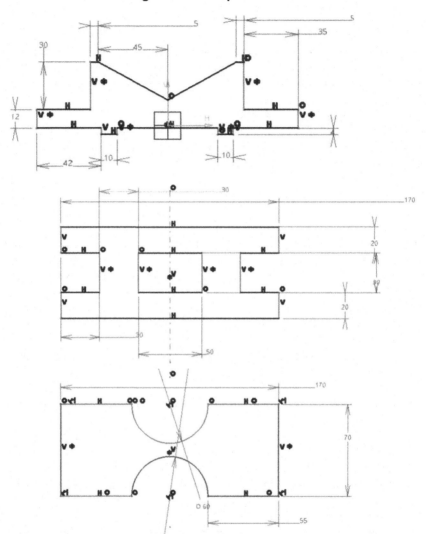

FIGURE 8.1   2D CAD model.

8.2.1 OVERVIEW OF THE BONDING PROCESS

There are three stages in the ultrasonic additive manufacturing bonding process.

In the first stage, the surfaces to be welded are drawn together by normal compression of the sonotrode. At the microscale, asperity tips are brought into contact and plastically deformed by the combined effect of normal stress caused by normal compression and interfacial shear stress caused by interfacial vibration.

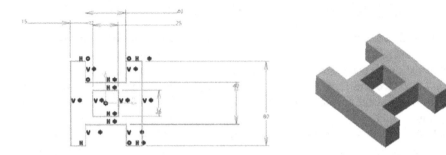

FIGURE 8.2 Extruding process.

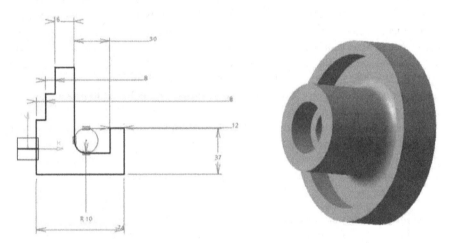

FIGURE 8.3 Revolving process.

Simultaneously cracks are produced in brittle surface oxides due to the difference in hardness between hard oxides and pure metals. The metal becomes even softer and the plastic regions are formed as the ultrasonic energy and the plastic and frictional heat dissipate into the material, making it easier to dissolve the surface oxides.

In the second stage, the metal-to-metal contact area increases and the interfacial voids are closed by the plastic flow as the welding cycle proceeds. Meanwhile, the broken oxides are carried by the metallic flow and dispersed to the edge of the welding zone.

In the third stage, a strong bond is formed across the interface where surface oxides are removed and metallic contacts are maintained. The bonds already formed are maintained by a plastic deformation that accommodates the interfacial vibration. The three stages of the bond cycle take place within very short time periods and are thus difficult to distinguish. For modeling purposes, the underlying assumption can be derived from the generalized three stages: Plastic deformation promotes bonding by dispersing surface oxides and contaminants, increasing contact areas of pure metal, and maintaining the bonds already formed.

8.2.2 BOND MECHANISMS

The bonding mechanism of the UAM has been studied for decades, but no uniform conclusion has been reached. As a bonding mechanism, metallurgical adhesion is endorsed by several researchers. The theory states that atom layers move across the bond interface and form "adhesive" bonds due to Van der Waals forces under intimate metal–metal contact. Intimate contact requires surface asperities and adjacent bulk material to undergo elasto-plastic deformation for the removal of surface oxides and the generation of metallic flows that fill the valleys between asperities. Diffusion through the welding interface is reported by some researchers on the basis of evidence of high strain rate plastic deformation observed. The high strain rate is expected to dramatically increase diffusion by rising vacancy concentrations within materials. In addition, the high vacancy concentration resulting from the high strain rate is expected to significantly lower the melting temperature of the material, allowing localized melting to occur. Recrystallization is also proposed as a source of bonding. The grains are observed to become finer in aluminum and copper after the UAM phase, suggesting the occurrence of recrystallization. Severe plastic deformation and temperature increase due to continuous input of ultrasonic energy are believed to provide the necessary conditions for recrystallization. Mechanical interlocking is reported by a few researchers who have studied the bonding of dissimilar materials as one material being soft and the other hard. Severe deformation of plastic is observed in soft material. In summary, plastic deformation is defined as the main factor regulating the bonding cycle. Specifically, it plays a vital role in the formation of bonds at all stages:

1) At the beginning of the bonding cycle, plastic deformation is observed in a thin layer of pure metal (~20 μm thick) below the surface oxides. The metallic flow helps to break up brittle oxides and disperse broken fragments.
2) When the oxides are removed and the pure metals are in contact, the plastic deformation of the asperity increases the metal-to-metal contact areas and the metallic flow closes the voids, resulting in a more complete, intimate foil contact and higher bonding quality.
3) When bonds are partially formed, layers of metal (20–60 μm thick) below bonded sites are believed to undergo plastic deformation in order to accommodate differential motion and to protect the bonds from breaking up. Moreover, although the exact bond mechanism is still subject to debate among researchers, plastic deformation is shown to improve bonding irrespective of the theories in use: Metal adhesion, diffusion, recrystallization, mechanical interlocking, and localized melting. As a result, it can be concluded that plastic deformation is a critical factor in the promotion of bond formation regardless of its causes.

8.3 COMMON FAULTS AND TROUBLESHOOTING

There are a lot of 3D printing troubleshooting issues that are bound to come up when we use machines. Some of the most common 3D-printing problems with the solutions in 3D printing are as follows.

8.3.1 THE PRINTER IS WORKING BUT NOTHING IS PRINTING

8.3.1.1 The Problem—Out of Filament

You have correctly designed the model in the slicing program, but nothing seems to happen; there's absolutely no printing going on. You have repeatedly sent a print to the printer, but all you get in return is a filament spit coming out of the nozzle, or may be the model is about to be printed, and out of nowhere the filament extrusion starts running, but the nozzle keeps going, printing nothing (Figure 8.4).

8.3.1.2 The Cause

While this problem is obvious on machines that have their filament reel in full view, like the PRUSA i3, there are machines that are not designed with exposed filament reels, such as the MOOZ, Ultimaker, and Robox, which will make it a little difficult to immediately detect the problem. These types of 3D printers either have their filaments encased or hidden at the back of the printer.

8.3.1.3 The Solution

The 3D-printing troubleshooting process here is relatively simple. No matter what type of 3D printer you use, all you have to do is remove the remaining file and load the new material. Check the filament reel for loading in another reel if there is no material at all.

8.3.2 NOZZLE IS TOO CLOSE TO THE PRINT BED

8.3.2.1 The Problem

You've loaded the filament perfectly, and nothing seems to be wrong with the print head, but there's no filament on the print surface (Figure 8.5).

FIGURE 8.4 Filament.

FIGURE 8.5   Nozzle too close to the print bed.

8.3.2.2   The Cause

Simply put, the nozzle and the print bed may be too close to each other. You may have accidentally turned your print bed a little away from your nozzle opening, thereby giving the melted filament little room to exit. The best case ideal situation with this issue is that your print will possibly miss its first few layers and the opportunities of it not sticking when the filament does not extrude will be high. In case of worst situation, the printer's hot-end will have a buffer of a molten filament which will certainly increase the chances of a blockage.

8.3.2.3   The Solution

There are two main ways to do this with 3D-printing troubleshooting:

  i. The Z-axis offset: This technique involves raising the nozzle a little bit. Most 3D printers have Z-axis offset settings in their system settings. With this setting, you can raise your nozzle a little high from the printing bed by setting a positive value to the Z-axis. Putting a negative value in the Z-axis offset setting will help you to resolve sticking issues, that is, if your prints don't stick to the print bed. Therefore, you need to make sure that when you set a value, it's not too high for your prints to stick to the bed.
  ii. Lower the bed: This is an alternative solution, as not all 3D printers allow the printing bed to be lowered. If printer allows this to happen, you can do it to fix this problem. It's a more upsetting fix, though, given that you're going to have to level and recalibrate your bed.

8.3.3   OVER-EXTRUSION

8.3.3.1   The Problem—Print Looks Droopy and Stringy

Simply put, this common 3D-printing problem simply means that the printer uses more material than it requires; thus, it creates more material than is required. Printed versions are likely to have excess materials on them (Figure 8.6).

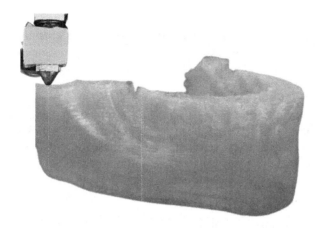

FIGURE 8.6   Over-extrusion.

FIGURE 8.7   Incomplete and messy infill.

8.3.3.2   The Cause

Essentially, the setting of the Flow or Extrusion multiplier in your slicing program is higher than usual.

8.3.3.3   The Solution

To troubleshoot this 3D-printing problem, go to your slicer program and test the settings of your Extrusion multiplier. Ensure that you have selected the correct value. If everything seems to be all right, then go to the Flow settings and decrease it.

8.3.4   INCOMPLETE AND MESSY INFILL

8.3.4.1   The Problem

Your print's internal structure is either broken or missing (Figure 8.7).

8.3.4.2 The Cause

To be frank, there are a number of reasons why the internal structure of your model can be broken or missing. While the most common one is that your slicing program has incorrect settings, a slightly blocked nozzle may also cause this problem.

8.3.4.3 The Solution

Take a look at the fill density—open your slicing software and take a look at the fill density. The best value is 20%; if the value is less than that, you're bound to encounter problems. However, if you're running large prints, you might want to raise this value to ensure that the model you're about to print is supported enough.

8.3.5 WARPING

8.3.5.1 The Problem—Bending

The printed model bends upward at the base until it no longer aligns with the printing platform. This results in the print being unplugged on the printing bed and the horizontal cracks developing in the upper parts of the printed models (Figure 8.8).

8.3.5.2 The Cause

Warping or cracking occurs because it is a natural characteristic of plastics. When your PLA or ABS filament begins to cool down, it gradually begins to contract. If the plastic cooling process happens too quickly, it results in warping or bending.

8.3.5.3 The Solution

There are different ways to troubleshoot this 3D printing issue.

1. Use a heated platform—it is the simplest solution; all you need to do is set the heated platform to a glass transition temperature (a temperature just below the melting point of the plastic). You can do this by using the slicer program

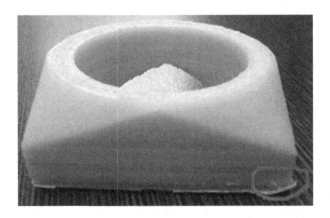

FIGURE 8.8 Warping.

and you can easily change the required filament temperature to the spool or wrapping of your filament. If you set the appropriate temperature, the first layer will remain flat on the surface.

2. If the calibration of the print platform is not right, you are bound to encounter this issue. So what kind of 3D-printing troubleshooting tip is going to work here? Well, all you have to do is level the print by checking the calibration again, make sure the bed is aligned, and the nozzle height is perfect.

8.3.6 Messy First Layer

8.3.6.1 The Problem

This is another common 3D-printing issue faced by many users. The first layers of printing are always troublesome. Problems are usually a non-stick print, or the bottom shell has an incorrect look due to unwanted lines. Also, instead of getting a fine detail on the bottom of your print, you find a blurry, congealed design that doesn't look like a surface design (Figure 8.9).

8.3.6.2 The Cause

The blurred and undefined detail on your print simply means that the temperature of the printing bed is too high. Unwanted lines arise as a result of the nozzle and the bed being too far away, blobs occur if the nozzle is too close to the bed. In addition, a non-stick print is the product of a bed that has not been properly leveled.

FIGURE 8.9 Messy first layer.

8.3.6.3 The Solution

Reduce the temperature of the bed and by lowering the temperature by five degrees at a time; continue to reduce the temperature until the desired adhesion result is achieved without losing any detail.

8.3.7 ELEPHANT'S FOOT

8.3.7.1 The Problem

Elephant's foot is a 3D-printing term that corresponds to the outward bulge of the base of the model. Simply put, it's when the printing bows or curves at the bottom (Figure 8.10).

8.3.7.2 The Cause

This usually happens when the weight of the model is pressing down on its base before it cools back to solid.

8.3.7.3 The Solution

Another 3D-printing troubleshooting tip you can use is to ensure that the base layers are cool enough to support the top structure. You need to make sure that the cooling is just the right level, since too much cooling causes the base layers to be warped. You might find this part tricky, but the easiest way to do this 3D-printing trouble-shooting process is to lower the printing platform's temperature by 5°C intervals to about plus or minus 25°C of the recommended temperature. If the bottom/top thickness is set to 0.6 mm, you can start the fan at a height slightly lower than that.

FIGURE 8.10 Elephant's foot.

8.3.8 Print Looks Deformed and Melted

8.3.8.1 The Problem

This issue is one of the most frequently asked questions in our 3D-printing FAQ. The filament has a highly resilient feature of all forms of misconfiguration, which makes it difficult for users to identify when the hot-end of their 3D printer is overheating. You just notice uneven layers, and when you take a closer look at the cabin, you'll see that the model is melted while you get something on the chimney that's close to the wax that's melted down the candle (Figure 8.11).

8.3.8.2 The Cause

The cause of the problem is an overheated hot-end. The temperature of your printer needs to be properly balanced to allow the filament to flow well and to allow it to solidify quickly. The balanced temperature will also make it possible to place the next layer on a more solid surface. However, before you adjust the temperature, make sure that the correct material is set up for the printer. If you check the material settings and everything looks fine, then all you might need to do is make a slight adjustment.

8.3.8.3 The Solution

 i. Check if your material settings are correct. Proper material temperature settings range from 180°C to 260°C.
 ii. Reduce the hot-end temperature of printer. This can be done with the hot-end temperature settings of software or printer. Reduce the temperature by five degrees Celsius, depending on the temperature recorded.

8.3.9 Snapped Filament

8.3.9.1 The Problem

Nothing comes out of the nozzle, but you can see it's full when you look at the filament spool and the feed tube seems to have some filament in it too. Bowden

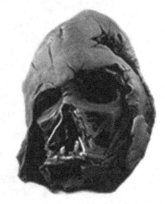

FIGURE 8.11 Deformed and melted print.

feed printers are unique to this problem; it is not usually found in direct feeds. That is because normally the filament is covered, so that you can't see it immediately (Figure 8.12).

8.3.9.2   The Cause

This problem can be caused by different factors but the most common one is cheap or old filament. Although it's true that most ABS and PLA filaments have a long lifetime, if held in bad weather, they may become brittle, once they go wrong, then when they are fed to the printer, no amount of change will make them right.

8.3.9.3   The Solution

i. Try to make use of another filament. If you find the same problem after the filament has been reloaded, try using another script. This would let you know if the brittle filament is actually causing the problem.
ii. Check the temperature and flow rate of your printer. This simply means that if the problem persists, check to see if the hot-end is at the right temperature and gets as hot as per suitable range. You should also check to see if the flow rate of the filament is no more than 100%.

FIGURE 8.12   Snapped filament.

8.3.10   Getting Cracks in Tall Objects

8.3.10.1   The Problem
You get cracks on the sides of your models, particularly when you're making taller models. This is considered to be one of the most surprising problems that can be experienced in 3D printing. That is because it usually appears in bigger prints, which typically happens while you're not looking (Figure 8.13).

8.3.10.2   The Cause
The reason for this problem is that the materials cool down faster in the higher layers of your print. The heat produced by the printing bed does not reach the upper pieces, making the upper layers less adhesive.

8.3.10.3   The Solution
Increase the extruder temperature by 10°C. Take a look at the side of the filament's box, you'll find the recommended temperatures for the hot-end. Try to keep your 3D printer's temperature setting within these values.

8.4   PROCESS DESIGN

The category of process design consists of research aimed at describing how the design is prepared for production. Liu and Rosen (2010) divide the process design

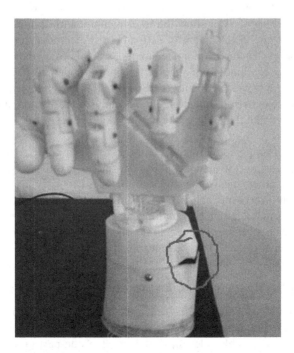

FIGURE 8.13   Cracks in objects.

into three steps: Component orientation, slicing, and process variable optimization. Jin et al. (2017) added the, steps generation, path generation and post-processing steps to the definition of AM process design.

The choices made will have an effect on what the best design of the part (and probably the system) would look like, and iterations are therefore important. The exact steps to be taken in the process design step could be argued, but all steps from design to manufacturing are discussed in this review. In this review, the three categories of support structure, manufacturing settings, and AM simulation of manufacturing are chosen.

8.4.1 CREATION END EVALUATION OF SUPPORT STRUCTURE

In the case of metal powder bed fusion, the AM support structure needs to be added in overhanging regions. The added support structure adds extra material to the manufacturing process, which adds production time, waste material, and post-processing time for the removal of the structure. Overhanging regions and the addition of support structures also create worse surface structures compared to other areas. In order to minimize the support required for the construction of a part, there are three alternatives: Optimization of the form and location of the support structure, optimization of the build direction, and modification of the design to make the part self-supporting. The support structure is directly linked to the choice of construction direction during development. Leutenecker-Twesiek et al. (2016) underline the importance of an early decision on part orientation for a product manufactured using AM. The purpose is to allow the implementation of design rules and guidelines that enable the development of self-supporting components and reduce the amount of support structure. Automated methods for choosing the best construction path to minimize the amount of support structure are described in (Strano et al., 2013a, 2013b) and (Zwier and Wits, 2016). Das et al. have merged various output goals for multi-objective optimization. In one analysis, the amount of support structure is minimized and the error of the part produced due to the effect of the stairs is minimized.

The amount of support structure and build-up time is minimized in another study. The creation of a support structure could be carried out in different ways and is often based on some mathematical algorithm that analyzes geometry in combination with the construction direction. Challenges in the design of a support structure include identifying areas that need support, reducing the amount of support, providing support with adequate mechanical properties (structural strength and heat dissipation), and providing support that is easy to remove.

8.4.2 ADDITIVE MANUFACTURING PREPARATION

The preparation step of the additive manufacturing process is to set up the system before the manufacturing process is carried out. This includes manufacturing settings that are highly relevant to the result of the part produced with AM. Nonetheless, it is difficult to provide an outline of the subject and there are no general guidelines. The types of settings that are available depend on which machine and which software is used to control the machine. Which settings are optimal also depends on the material, the geometry, and whether the other components are made at the same time. The manufacturing settings could be divided into four types: Energy-related, scan-related, powder-related, and temperature-related. Energy-related settings include energy source power, spot size, pulse duration, and pulse frequency. Scan-related settings include scanning speed, scanning spacing, and scanning pattern. Powder-related settings are connected to the substance used and include the shape and size of the particle as well as how the powder is dispersed and the thickness of the coating used. Temperature-related parameters include the temperature of the powder bed, the feeder, and the temperature consistency. Both parameters are highly dependent on each other, and changing one would also affect the other parameters. Many additive manufacturing companies use standard settings for various materials and machines, making it difficult for the design engineer to change the settings. Instead, the most common way is to change the design if there are manufacturing errors. In a perfect future, the design phase would involve a feedback loop where geometry and manufacturing settings are managed together to optimize the device as a whole.

8.4.3 Validation of Build Time and Cost

The estimation of the construction time of the part is crucial in order to be able to calculate the cost of production. In the area of cost simulations for additive manufacturing, Costabile et al. (2017) carried out a comprehensive review of the different research studies and concluded that, no matter which AM technique is used, the cost model looks similar. Several different models are highlighted and presented in more detail. Chiu and Lin (2016) investigated the possibility of producing a simulated business case to determine whether or not a product is suitable for conversion to AM, including a cost model combined with design for additive manufacturing to optimize costs based on design and production techniques.

8.4.4 Additive Manufacturing Simulation

Numerous methods have been used in both research and commercial applications to model the additive manufacturing process, resulting in efficiency, surface quality, and dimensional accuracy of the final component. Bikas et al. (2016) divided the simulation approaches into three categories—analytical, numerical, and empirical—on the basis of which principle is used. Analytical simulation models are based on physical laws that have the advantage that they can be easily adapted to different processes, machines, and machine settings.

However, they are limited by the initial observations that need to be made in order to apply the laws of physics. The analytical method is focused on observation and is therefore reliable for the exact set-up of the test, but is more difficult to adjust to other devices and set-ups. The numerical approaches attempt to combine the two other methods and start with an analytical model that is combined with a numerical model.

8.5   LOW COST, RAPID DEPLOYMENT WIRELESS PATIENT MONITORING SYSTEM DEVELOPED WITH ADDITIVE MANUFACTURING EQUIPMENT: CASE STUDY

Mr. Summet Gattewar is the Director of Pye Technologies India and he has initiated this venture as Pye Technologies India, which is a Nagpur-based startup that has indigenously developed a low-cost wearable device to meet the patient monitoring requirements of the COVID-19 pandemic in India. COVID-19 in India is going to be a very different challenge than China, Italy, France, and the United States owing to its higher population density and lower healthcare infrastructure. The Government of India is taking extreme steps to control the spread of coro
navirus in India. However, if the number of COVID-19 patients in India increases, there will be a need to set up temporary ICU wards and isolation zones. Efforts to establish isolation wards have already begun, but the patient monitoring needs of these wards are difficult to meet given the high cost of monitoring equipment ranging from forty thousand to two lacs of rupees per bed, and the large quantity required.

Health monitoring systems have rapidly evolved during the past two decades and have the potential to change the way health care is currently delivered. Although smart health monitoring systems automate patient monitoring tasks and thereby improve the patient workflow management, their efficiency in clinical settings is still debatable. This case study presents a review of smart health monitoring systems and an overview of their design and modeling with 3D printing.

For the patient monitoring requirements of COVID-19 in India, a low cost device—a wrist band—is designed and developed using 3D printing. This watch monitors three medical parameters viz. blood oxygen saturation, heart rate, and body temperature. The band communicates wirelessly to a central station which captures and stores data. This data can be viewed and analyzed remotely with the help of cloud platform. Multiple bands can be monitored centrally on site using our dashboard which opens on any standard desktop or mobile device. Doctors can monitor patients' parameters in real time, view trends, and get notifications for cases which need attention. The complete system has also been tested for the cloud.

The following hardware, which has been indigenously developed at Pye Technologies India, costs around Rs. 10,000 per patient. The system is based on wireless communication which ensures high scalability and rapid deployment for creating temporary wards or isolation zones equipped with real time patient moni-toring capability (Figures 8.14 to 8.16).

This prototype was developed on low-cost in-house developed CNC machines which also utilizes the various parts printed with FDM 3D printing (Figures 8.17 and 8.18).

The device was tested at New Era Hospital and Research Institute, Lakadganj, Nagpur under the supervision of Dr. Anand Sancheti. The device was found to be acceptable and has been recommended by doctors on the basis of the field tests

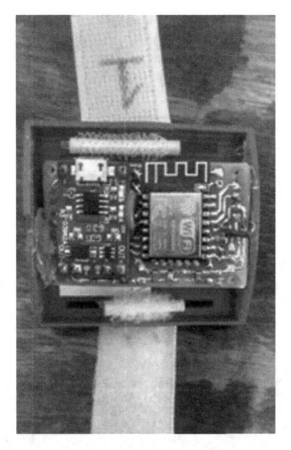

FIGURE 8.14 Case Study picture 8.1. (Courtesy: Pye Technologies, India.)

FIGURE 8.15 Case Study picture 8.2. (Courtesy: Pye Technologies, India.)

FIGURE 8.16   Case Study picture 8.3. (Courtesy: Pye Technologies, India.)

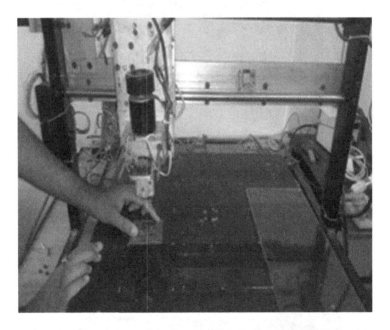

FIGURE 8.17   Case Study picture 8.4. (Courtesy: Pye Technologies, India.)

carried out. Furthermore, a critical analysis of the efficiency, clinical acceptability, strategies, and recommendations on improving current health monitoring systems has been tested. The main aim is to review the current state of the art monitoring systems and to perform extensive and in-depth analysis of the findings in the area of smart health monitoring systems developed through 3D printing.

FIGURE 8.18   Case Study picture 8.6.

8.5.1   CONCLUSION

Finally, major advances in the system design level and current issues facing health care providers have been investigated, and the potential challenges to the health monitoring field have been identified and compared to other similar systems. The results demonstrated by Mr. Summet Gattewar, along with the authors of the book, suggested that the proposed system could be comparable to medical grade devices. The patent on a low-cost, rapid deployment wireless patient monitoring system has been filed and mass production is expected to start to sustain the effects of COVID-19 in India.

8.6   EXERCISES

1. Name the design and process parameters of process equipment and briefly explain seven distinct AM processes.
2. Explain the designing for 3D printing and factors to be considered when designing for 3D printing.
3. Explain in detail the governing bonding mechanism.
4. Discuss the common faults and troubleshooting in 3D printing.

8.7 MULTIPLE-CHOICE QUESTIONS

1. Parts produced using a 3D printer must be designed using some kind of
   a) CATIA software
   b) ProE software
   c) CAD software
   d) Manual design
   Ans: (c)

2. Elephant's foot is a 3D-printing term that corresponds to the _____
   bulge of the base of the model.
   a) Outward
   b) Inward
   c) Curved
   d) None of the above
   Ans: (a)

3. Choose the remedy to reduce the chance of a messy first layer.
   a) Increase the temperature of the bed
   b) Reduce the temperature of the bed
   c) Increase the temperature of the nozzle
   d) Reduce the temperature of the nozzle
   Ans: (b)

4. Select the solution to reduce the chance of getting cracks in tall objects.
   a) Decrease the extruder temperature by 10°C
   b) Increase the bed temperature by 10°C
   c) Decrease the extruder temperature by 10°C
   d) Increase the extruder temperature by 10°C
   Ans: (d)

5. If I have a model with hands that stand out for the sides, what do I have to
   turn on to print it?
   a) Raft
   b) Structure
   c) Supports
   d) Fixture
   Ans: (c)

6. If I want to quickly make a cup and only use a spline/line to create it. Which
   tool do I use to make it 3D?
   a) Sweep
   b) Loft
   c) Extrude
   d) Revolve
   Ans: (d)

9 Post-Processing

9.1 INTRODUCTION

AM parts also require post-processing to enhance surface finishing and mechanical properties. For metals, the dimensional precision of additive processes is not yet adequate to produce a component that can reach tight tolerances without further processing. Many AM processes need post-processing to prepare the component for its intended shape, fit, and function after partial construction. Depending upon the AM procedure, the explanation for post-processing varies. This chapter will concentrate on post-processing techniques for convenience purposes which are used to enhance components or address AM limitations. These include:

- Support material removal
- Surface texture improvements
- Accuracy improvements
- Asthetic improvements
- Easy to use as a pattern
- Improvements to properties using non-thermal techniques
- Enhancements in property using thermal techniques

There are a number of process parameters that can be varied to influence the physical, mechanical, thermal properties, cost, speed, and quality of an AM component. The combination of process parameters coupled with post-processing directly influences the microstructure (e.g. grain size, porosity, cracking density), and mechanical properties, such as tensile, fatigue, and creep properties.

The willingness of numerous AM professionals to conduct post-processing is one of the most distinguishing characteristics of competitive service providers. Companies that can effectively and accurately post-process parts to the customer's specifications will often charge a premium for their services, whereas companies that compete solely on price can compromise post-processing quality in order to reduce costs.

9.2 SUPPORT MATERIAL REMOVAL

Support removal is the most common type of post-processing in AM. Help materials can be loosely divided into two categories:

(a) Material that surrounds the part as a naturally occurring by-product of the construction process (natural supports) and
(b) Rigid structures that are designed and built to support, restrain, or attach the part that is being built to the building platform (synthetic supports)

9.2.1 NATURAL SUPPORT POST-PROCESSING

Until usage, the part must be separated from the surrounding material in processes where the part being manufactured is entirely encapsulated in the building material.

Processes which provide natural support are primarily processes based on powder and surface. In particular, both processes of powder bed fusion (PBF) and binder jetting require removal of the part from the loose powder surrounding the part; and processes of bond-then-form sheet metal lamination require removal of the encapsulating sheet material. In polymer PBF processes, it is usually important to allow the component to pass through a cool-down stage after the component has been constructed. The part will remain inside the powder to reduce the distortion of the component due to non-uniform cooling. The cool-down time depends on the building material and the scale of the part(s). When cool-down is complete, many methods are used to separate the part(s) from the surrounding loose powder. The entire structure (made of loose powder and fused parts) is usually removed from the machine as a block and transferred to the "breakout" station where the pieces are separated manually from the powdered material surrounding it. Brushes, compressed air, and light bead blasting are widely used to extract loosely bound powder; also, woodworking devices and dental cleaning tools are generally used to extract powders that have been sintered to the surface or powder that has been trapped in small channels or features. Internal cavities and hollow spaces can be difficult to clean, and can take considerable time after processing.

Natural support removal methods for binder jetting processes are the same as those used for PBF, except for extended cool-down periods. Components made from binder jetting are in most cases brittle out of the system. Therefore, the pieces must be treated with caution before the pieces have been strengthened by infiltration. It also applies to post-infiltration PBF materials, such as other elastomeric materials, investment casting materials from polystyrene, and green sections from metal and ceramics. More recently, automated processes for extracting material have been developed, which can be stand-alone or incorporated into the build chamber. One of the first ZCorp (now 3D Systems) binder jetting machines with this capability is shown in Figure 9.1. Several metal PBF machine manufacturers have also begun to integrate semi-automated powder removal techniques into their devices. Current trends suggest that a number of future PBF and binder jetting machines may implement some form of automated powder removal after part completion.

Bond-then-form sheet lamination processes, such as Mcor machines, also require the removal of natural support material prior to use. When using complex geometries with overhanging components, internal cavities, channels, or fine features, it can take time to remove the support. When enclosed cavities or channels are formed, de-laminating the model at a particular z-height is often necessary to gain access to the internal de-cube feature; and then re-gluing it after removing excess support materials. An example of LOM de-cubing operation is shown in Figure 9.2.

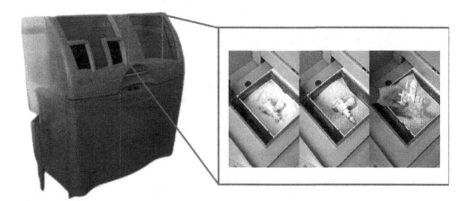

FIGURE 9.1 Automated powder removal using vibratory and vacuum assist in a ZCorp 450 machine.

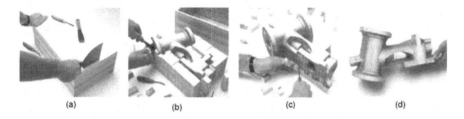

FIGURE 9.2 LOM support removal process (de-cubing), showing: (a) The finished block of material; (b) Removal of cubes far from the part; (c) Removal of cubes directly adjacent to the part; (d) The finished product.

9.2.2 SYNTHETIC SUPPORT REMOVAL

Processes that don't support components necessarily need synthetic support for over-hanging features. For certain cases, for example, synthetic supports are also needed to withstand distortion when using PBF techniques for metals. Synthetic supports may be made from a material of construction or a material of secondary significance. The production of secondary support materials has been a critical step in simplifying the removal of composite supports, as these components are either engineered to be softer, dissolve in a liquid solution, or melt at a lower temperature than the build-ing material. The part's orientation toward the primary axis of the building greatly affects support generation and removal. For example, if a thin part is laid flat, the amount of support material used may considerably exceed the amount of build mate-rial (see Figure 9.3).

Support orientation often affects the surface finish of the part, as the removal of the support usually leaves "witness marks" (small bumps or divots) where the sup-ports were attached. In addition, the use of support in regions with small features may cause these features to be broken when the support is removed. The orientation and location of the supports is therefore a key factor in the achievement of desirable finished part characteristics in many processes.

FIGURE 9.3 The flat FDM-produced aerospace part white build material is ABS plastic and the black material is the water-soluble WaterWorks™ support material.

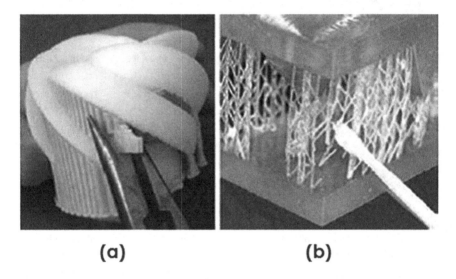

FIGURE 9.4 Breakaway support removal for: (a) An FDM part; (b) An SLA part.

9.2.2.1 Supports Made from the Build Material

All material extrusion, material jetting, and vat photopolymerization processes need support for overhanging structures and for connecting the component to the construction platform. Since these processes are used mainly for polymer parts, the low strength of the supports enables them to be removed manually. These forms of support are also commonly referred to as support for breakaways. Removing the supports from the downward-facing features leaves traces on witness where the supports were connected. It may include subsequent sanding and polishing of these surfaces. Figure 9.4 demonstrates the breakaway removal techniques for parts made from extrusion material and vat photopolymerization techniques.

PBF and DED processes for metals and ceramics often typically require supporting materials. An example of a dental framework, designed to prevent the removal of support from the vital surfaces, is shown in Figure 9.5.

For these methods, metal supports are often too strong to be extracted by hand; thus, the use of milling, bandsaws, cutting blades, wire-EDMs, and other metal

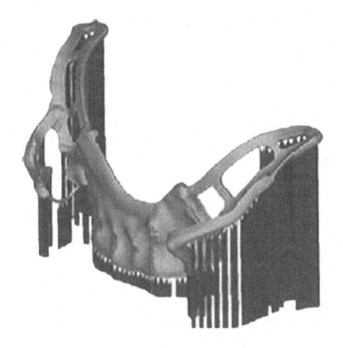

FIGURE 9.5   SLM dental framework.

cutting techniques is commonly used. The components made by electron beam melting have fewer supports than those made by metal laser sintering, as EBM keeps the part at high temperatures throughout the construction process and induces less residual stress.

9.2.2.2 Supports Made from Secondary Materials

Over the years, a number of secondary support materials have been developed to reduce the labor-intensive manual removal of support materials. Two of the first techniques to use secondary supporting materials were the Cubital layer-wise vat photopolymerization process and the Solidscape surface jetting process.

The use of wax support materials made it possible for the support block/build to be placed in a warm water bath; therefore, the melting or dissolving of the wax produces the final parts. Since then, secondary supports have become common commercially in material extrusion (Figure 9.3) and material jetting processes. Secondary support for form-then-bond sheet metal lamination and DED processes in research environments has also been demonstrated. The most common secondary support materials for polymers are polymer materials that can be melted and/or dissolved in a water-based solvent. Water can be jetted or ultrasonically vibrated to accelerate the recovery process. For metals, lower melting-temperature alloys or alloys are the most common secondary support materials that can be chemically dissolved in a solvent (in this case, the solvent does not influence the build material).

9.3 SURFACE TEXTURE IMPROVEMENTS

AM parts have common surface texture characteristics that may need to be changed for esthetic or performance reasons. Popular undesired surface texture features include: Stairway, powder adhesion, filling patterns from extrusion material or DED systems, and testimonial marks from support material removal. Stair-stepping is a key issue in layered production, although a thin layer thickness may be chosen to minimize error at the expense of construction time. Powder adhesion is a key feature of binder jetting, PBF, and powder-based DED processes.

The amount of powder adhesion can be regulated to some degree by changing the orientation of the part, the morphology of the powder, and the thermal control technique (e.g. by modifying the scan pattern).

The kind of post-processing used to improve the texture of the surface depends on the desired finish. If a matte surface finish is required, a simple surface bead blasting will help even the texture of the wall, remove sharp edges from the stairs, and give an overall matte appearance. If a smooth or polished finish is required, wet or dry sanding and hand polishing shall be carried out. In certain situations, it is best to paint the surface before sanding or polishing (e.g. with cyanoacrylate or sealant). Painting the surface has the dual benefit of sealing the porosity and smoothing the stair-step effect through viscous forces, making sanding and polishing simpler and more efficient.

Many automated techniques have been explored to improve surface texture. Two of the most commonly used are tumbling for external characteristics and abrasive flow machining, mainly for internal functions. Such processes have been shown to be smooth on the surface, but at the cost of small feature resolution, good corner retention, and accuracy.

9.4 ACCURACY IMPROVEMENTS

Between AM processes, there is a large variety of accuracy capabilities. Several processes are capable of submicron tolerances, while others have precision of around 1 mm. Usually, the higher the build volume and faster the build speed, the worse the precision. It is especially evident, for example, in directed energy deposition processes where the slowest and most reliable DED processes have accuracy of a few microns; while larger bulk deposition machines have accuracy of several millimeters.

9.4.1 SOURCES OF INACCURACY

Process-dependent errors affect the X–Y plane accuracy differently than the Z-axis accuracy. These errors arise from the positioning and indexing limitations of particular system architectures, the lack of closed-loop process monitoring and control techniques, and/or issues that are fundamental to the volumetric rate of material addition (such as melt pool or droplet size). However, for many operations, accuracy depends heavily on the operator's abilities. Future improvements in AM accuracy will require fully automatic real-time monitoring and control systems, rather than relying on expert operators as feedback mechanisms. Additive plus subtractive processing integration is another method for improving process accuracy.

Material-dependent anomalies also play a role in precision, including shrinkage and residual stress-induced distortion. Reproducible shrinkage and distortion can be compensated by scaling the CAD model; however, predictive capabilities are currently not accurate enough to fully understand and compensate for variations in shrinkage and residual stress depending on the scan pattern or geometry. To improve these predictive capabilities, quantitative understanding of the effects of process parameters, design style, component orientation, support structures, and other factors are required on the magnitude of shrinkage, residual stress, and distortion. Furthermore, additional material must be applied to the essential features for parts that require a high degree of precision, which are then extracted by milling or other subtractive methods to achieve the desired accuracy. To satisfy the needs of applications where the advantages of AM are required with the accuracy of the CNC machined component, a rigorous strategy may be implemented to achieve this accuracy.

One such technique involves pre-processing the STL file to compensate for inaccuracies that are followed by the finishing of the final portion. The following sections describe the steps that need to be considered when developing a detailed finishing machining strategy.

9.4.2 MODEL PRE-PROCESSING TO COMPENSATE FOR INACCURACY

The location of the part within the build chamber and the orientation will affect the precision of the part, the surface finish, and the construction time for many AM processes. Then, translation and rotation operations are applied to the original model to improve the component's position and orientation. During the AM time, shrinkage also occurs. Shrinkage often occurs during post-process furnace operations required for indirect processing of green metal or ceramic pieces. Pre-process manipulation of the STL model will make it possible to use a scale factor to compensate for the average shrinkage of the process chain. Nonetheless, when the average shrinkage is compensated, there will always be some features that decrease slightly more or less than the average (shrinkage variation).

In order to compensate for shrinkage variance, if the maximum shrinkage value is used, the ribs and similar features will always be at least as large as the desired geometry. But the channels and holes will be too large. Therefore the easy use of the greater shrinkage value is not an acceptable solution. To ensure that there is enough material left on the surface to be machined, the original model must be replaced with "skin." This removal of skin, such that material is left to the machine everywhere, can be called making the "steel-safe" component.

Several studies have shown that differences in shrinkage rely on geometry, even when applying the same post-processing parameters for AM or furnace. Shrinkage variance compensation also requires offsetting of the original model to ensure that even the features with the highest shrinkage rates and all channels and holes are steel-safe. There are two principal methods for applying the skin to the part's surface. The first is offsetting the surface, and then recalculating all surface intersections.

Although most common, this technique has many disadvantages for triangular facets shaped STL files. In response to these drawbacks, an algorithm developed to offset all the individual vertices of the STL file using the normal vector information for the connected triangles, and then to reconstruct the triangles using the new vertex values, has been developed. In the STL file, each vertex is commonly shared by several triangles whose normal vector unit is different. When offsetting the vertices of a graph, the new value of each vertex is determined by the normal unit values of its related triangles.

Suppose V offset is the unit vector from the original position to the new position of the vertex to be moved, and $N_1, N_2......N_n$ is the unit of the normal vectors of the triangles which share that vertex; V offset can be determined by the weighted mean of

$$\overline{V}_{offset} = \sum_{i=1}^{n} W_i \overline{N}_i \qquad (9.1)$$

where W_i are coefficients whose values are determined to satisfy the equation

$$\overline{V}_{offset} = \overline{N} = 1\,(i-1,2,...n) \qquad (9.2)$$

After solving for V offset, the new position P_{new} of the vertex is given by the equation

$$P_{new} = P_{original} + \overline{V}_{offset} * d_{offset} \qquad (9.3)$$

Where d_{offset} is the offset dimension set by the user.

The procedure referred to above is repeated until the new position values are calculated for all vertices. The model is then reconstructed using the new information on the triangle. Thus, in order to use this offset methodology, it is only necessary to enter a d_{offset} value that is the same as the most anticipated shrinkage variation. In practical terms, the d_{offset} should be set at two or three times the absolute standard deviation of the shrinkage measured for a particular machine/material combination.

9.4.3 MACHINING STRATEGY

Machining strategy plays an important role in the finishing of AM parts and tools. Considering both precision and machine efficiency, adaptive surface grinding, plus hole drilling and sharp edge grinding can meet the needs of most parts.

9.4.3.1 Adaptive Raster Milling

For milling operations, the stepover distance between adjacent tool paths is a very important parameter that controls the precision and surface quality of the machine. It is known that better accuracy and surface quality require a smaller stepover distance. Normally, the height of the material left after the model has been machined is used as a measure of the surface quality.

Figure 9.6 shows the triangle face being machined by the ball end mill. The relationship between the cusp height h, the cutter radius r, the stepover distance d, and the incline angle α is given in the following equation:

$$d = 2.0\sqrt{r^2 - (r-h)^2} \, \cos\alpha \qquad (9.4)$$

α is determined by the triangle surface normal and stepover direction. Suppose N triangle is the unit normal vector of the triangle surface, and $N_{Stepover}$ is the unit vector along stepover direction, then

$$\cos\left(\frac{\pi}{2} - \alpha\right) = \sin\pm = \left|\overline{N}_{Triangle} \cdot \overline{N}_{Stepover}\right| \qquad (9.5)$$

From (9.4) and (9.5), the following equation for stepover distance is derived,

$$d = 2.0\sqrt{h(2r-h)\left(1 - (\overline{N}_{Triangle} \cdot \overline{N}_{Stepover})^2\right)} \qquad (9.6)$$

The cutter radius and milling direction are the same for all triangle surfaces when machining the pattern. If the maximum cusp height h is defined by the user, d is only connected to the usual vector triangle. For surfaces with different normal vectors, the distance obtained will be different.

If a constant stepover distance is used to ensure a consistent machining tolerance, the minimum measured d for the whole part should be used. However, the use of minimum stepover distances will lead to longer programs and machining times. An adaptive stepover distance for milling operations based on local geometry should therefore be used to allow both precision and machine performance. This means that stepover distances for each just completed tool pass are dynamically determined,

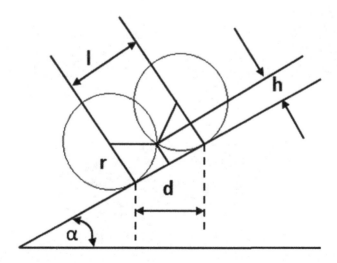

FIGURE 9.6 Illustration for determining stepover distance.

using the maximum cusp height to evaluate the stepover distance for the next pass. An example of how to use this type of algorithm for generating tool paths is shown in Figure 9.7. As can be shown, the tool paths are more closely spaced for tool paths that move through a high angle region; while tool paths that only cross fairly flat regions are broadly spaced.

9.4.3.2 Sharp Edge Contour Machining

Sharp edges are often curves of intersection between features and surfaces. Normally, the critical dimensions are defined by these edges. Using raster milling, edges parallel to the milling path can be skipped, resulting in significant errors. As shown in Figure 9.8, if the stepover distance d is used to machine a part with a slot width W, even if the CNC machine is perfectly aligned (i.e. ignoring machine positioning errors), the slot width error will be at least,

$$W_{error} = 2d - \delta_1 - \delta_2 \tag{9.7}$$

where $\delta 1$, $\delta 2$ represent the offset between the actual and desired edge location.

When $\delta 1$, $\delta 2$ become 0, $W_{error} = 2d$. This means that the possible maximum error for a slot using raster milling is approximately two times the stepover distance.

For complicated edges that are not parallel to the milling path, raster milling is ineffective for producing smooth edges, as the edge will have a step-step appearance, with a step size equal to the local step-step width, d. Therefore, after raster milling, it is advantageous to run a machining pass along the sharp edges (contours) of the component. In order to be machined along sharp edges, the STL model must

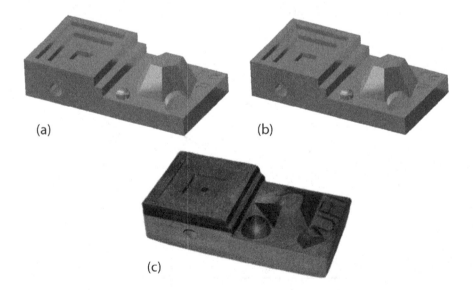

(a)   (b)

(c)

FIGURE 9.7 Finish machining using adaptive raster milling of a copper-filled polyamide part made using polymer laser sintering: (a) CAD model, (b) tool paths, (c) machined part.

first identify all sharp edges. Normal vector information for each triangle is used to check the edge property. The angle between the normal vectors of the two adjacent triangles is calculated. When this angle is greater than the user-specific angle, the edge shared by these two triangles will be identified as a sharp point. All triangle edges are checked in this way to generate a sharp edge list. Hidden edges and redundant paths of the tools are eliminated before measuring the paths of the tools. The x, y orientation of the endmill is obtained by offsetting the edges by the cutter radius. The value z is calculated by determining the intersection with the 3D model and finding the maximum value z corresponding. The sharp edges can be identified and easily machined using this approach. Figure 9.9 shows the part of Figure 9.7 marked with sharp edge contour paths (Figures 9.8 and 9.9).

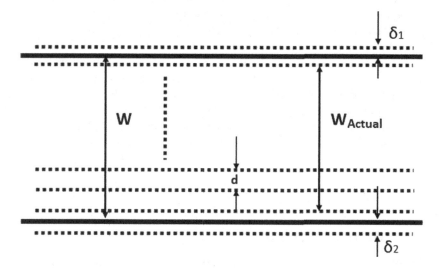

FIGURE 9.8 Influence of stepover distance on dimensional accuracy.

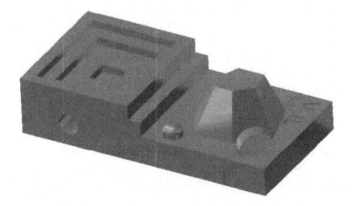

FIGURE 9.9 Sharp edge contours identified for milling.

9.4.3.3 Hole Drilling

Circular holes are common features of the components and tools. It is expensive to use the milling tools to create holes and the circularity of the holes is low. A machining strategy for the identification and drilling of holes is therefore preferable. The most difficult aspect is to identify holes in the STL or AMF file, as the 3D geometry is represented by a set of unordered planar triangular facets (and thus all the feature information is lost). The curve of the intersection between the hole and the surface is usually a closed loop. Using this information a hole recognition algorithm begins by finding all the closed loops consisting of the pattern's sharp edges. The intersection curves between the holes and the surface are not actually such closed loops, and a set of hole-checking rules are used to eliminate loops that do not suit the drilled holes.

The remaining loops and their normal vectors on the surface are used to determine the drilling process diameter, axis orientation, and depth. Tool paths can be generated automatically from this knowledge.

Thus, by pre-processing the STL file using the shrinkage and offset value of the surface and then post-processing the part using adaptive raster milling, contour grinding, and hole drilling, an exact part can be made. In many cases, however, this type of comprehensive strategy is not essential. For example, for a complex component where only one or two properties need to be rendered correctly, the component may be pre-processed using the average shrinkage value as a scaling factor and a skin can only be applied to the critical characteristics. Upon AM component development, these essential characteristics could then be machined manually, leaving the other features as they are. The finish machining approach adopted would thus rely heavily on the application requirements and part-specific design specifications.

9.5 ESTHETIC IMPROVEMENTS

AM is often used to make parts that are shown for esthetic or artistic purposes or used as marketing tools. In these and similar cases, the esthetics of the part are of critical importance for its final application. Often the desired esthetic improvement is solely related to the surface finish. In this case, the options for post-processing are discussed in Section 9.2. For certain cases, a variation for surface texture may be required between one area and another (this is always the case with jewelry). In this case, it only involves the finishing of selected surfaces (for example, the cover art for this book). In cases where the AM component's color is not of good consistency, certain methods can be utilized to improve the product's esthetics.

By simply immersing the item in a dye of the correct color, some types of AM parts can be effectively colored. This method is especially effective for parts made from powder beds, because these parts' inherent porosity results in effective absorption. The part may need to be sealed before painting, if painting is necessary. In these cases, car paints are very successful.

Chromium plating is another esthetic enhancement (which also strengthens the part and improves wear resistance). Figure 9.10 shows the stereolithography part before and after chromium plating. Several materials have been electrolyzed to AM

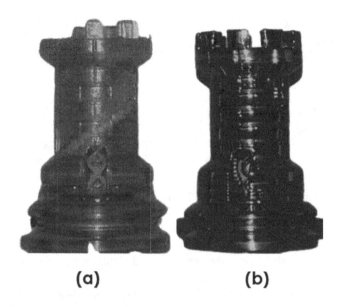

FIGURE 9.10 Stereolithography part (a) before and (b) after chrome plating.

parts, including Ni, Cu, and other coatings. In some cases, these coatings are thick enough that, in addition to esthetic enhancements, the parts are strong enough to be used as injection molding tools or as EDM electrodes.

9.6 PREPARATION FOR USE AS A PATTERN

Often AM-based parts are intended as investment casting patterns, sand casting, room temperature vulcanization (RTV) molding, spray metal deposition, or other pattern replication processes. Using the AM pattern in the casting process is in many cases the cheapest way to use AM to manufacture a metal component, as many of the metal-based AM processes are still expensive to own and run. The AM pattern's accuracy and surface finish will directly affect the accuracy of the end part and surface finish. Therefore great care must be taken to ensure that the pattern in the final part has the desired consistency and surface finish. The pattern must also be scaled to account for any shrinkage that occurs in the replication stage of the pattern.

9.6.1 INVESTMENT CASTING PATTERNS

The AM pattern will be used during manufacturing, in the case of investment casting. The residue left in the mold when the pattern is melted or burned out is in this case undesirable. Any sealant used to flatten the surface during pattern preparation should be carefully selected so as not to unintentionally produce unnecessary residues.

The AM parts can be printed on the casting tree or added manually to the casting tree after AM. Figure 9.11 displays rings made from a material jetting device.

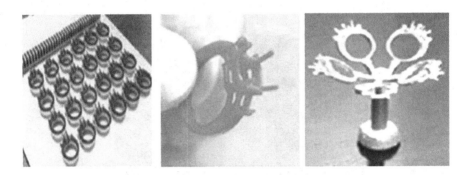

FIGURE 9.11   Rings for investment casting, made using a ProJet® CPX 3D printer.

In the first picture, a collection of rings is shown on the construction platform; each ring is supported by a secondary support material in white. A close-up of the ring pattern is shown in the second picture. The third picture shows the metal rings still attached to the cast tree. In this instance, after AM, but before casting, the rings were added to the tree. When using Quickcast style stereolithography, hollow, truss-filled shell patterns must be drained of liquid prior to investment. The hole(s) used for drainage shall be covered to prevent investment from reaching the interior of the pattern. Since thermosets are photopolymer materials, they must be burned rather than melted out of the investment.

The resulting component is brittle and fragile when using powdered materials as investment casting patterns such as polystyrene from a polymer laser sintering process, or starch from a binder jetting process. To seal and reinforce the part for the investment process, the part is pre-investment infiltrated with an investment casting wax.

9.6.2   Sand Casting Patterns

Binder jetting and PBF processes can be used to directly create sand mold cores and cavities using a thermoset binder to bind sand in the desired shape. One benefit of these direct approaches is that complex geometry cores can be produced that would be very difficult to produce by any other method, as illustrated in Figure 9.12. To prepare AM sand casting patterns for casting, loose powder is removed, and the pattern is heated to complete the thermoset binder cross-linking and eliminate by-products of moisture and steam. For certain cases, additional binders are applied before heat to the pattern to increase handling resistance. When the pattern is heat treated, the corresponding core(s) and/or cavity are assembled and hot metal is poured into the mold. Sand patterns are collected after refrigeration using tools and bead blasting.

In addition to directly creating sand casting cores and cavities, AM can be used to create parts that are used instead of traditional wood or metal patterns from which sand casting molds are made. In this case, the AM part is built as one or more portions of the component to be cast, separated along the parting line. The part is put in a jar, the sand mixed with the binder is poured around the part, and the sand is

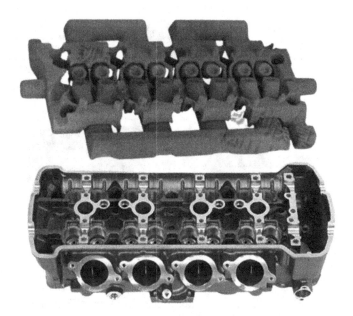

FIGURE 9.12 Sand casting pattern for a cylinder head of a V6, 24-valve car engine (left) during loose powder removal and (right) pattern prepared for casting alongside a finished casting.

compressed (pounded) so that the binder keeps the sand together. The box is disassembled, the sand mold is removed from the box, and the pattern is removed from the mold. The mold is then reassembled with its complementary mold half and core(s) and the molten metal is then poured into the pipe.

9.6.3   OTHER PATTERN REPLICATION METHODS

Since the late 1980s, there have been several pattern replication methods used to turn the weak "rapid prototypes" of those days into pieces with useful material properties. While the number of AM technologies has increased and the quality of the products they can manufacture has greatly improved, these processes of duplication are seeing less use as people choose to directly produce a functional component, if possible. Even with the multiplication of AM technologies and materials, however, pattern replication processes are commonly used by service offices and companies that need parts from different materials that cannot be processed directly in AM.

The most common pattern replication methods are probably RTV molding or silicone rubber molding. In RTV molding, as shown in Figure 9.13, the AM pattern is given visual markers (such as the use of colored tape) to show the parting line positions for the disassembly of the mold; the runners, risers, and gates are added; the model is placed in the mold box; and the rubber-like material is poured around the mold to encapsulate it. After cross-linking, the solid, translucent rubber mold is removed from the mold box, the rubber mold is cut into pieces by the parting line

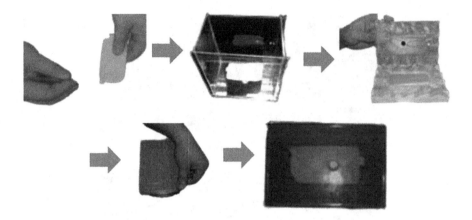

FIGURE 9.13   RTV molding process steps.

marker, and the pattern is removed from the mold. The mold is reassembled and held together in a box, or by placing rubber bands around the mold to complete the replication process, and the molten material is poured into the mold and allowed to solidify. The mold is opened after solidification, the part is removed, and the process is repeated until there is a sufficient number of parts. This technique can be used to produce 10s or 100s of the same pieces using a single pattern. If the part formed in the RTV mold is a wax template then a metal part can be produced using it in an investment or plaster casting process.

Thus a single AM design can be repeated to a large number of metal parts at relatively modest cost by combining RTV molding and investment casting. Metal spraying processes were also used to reproduce geometry from part AM to part metal. In the case of a metal spray, the metal part is repeated only by one side of the pattern. It is most commonly used for tooling or sections where one side contains all the complexity of the geometry, and the rest of the tool or component consists of smooth edges. The AM pattern may be repeated using spray metal or electroless deposition processes to create an injection molding core or cavity which can then be used to mold other parts.

9.7   PROPERTY ENHANCEMENTS USING NON-THERMAL TECHNIQUES

Powder-based and extrusion-based processes often generate porous structures. In certain cases, this porosity can be infiltrated by a higher strength substance such as cyanoacrylate (Super Glue). Proprietary methods and materials have also been developed to enhance the strength, ductility, heat deflection, flammability resistance, EMI shielding, or other properties of AM parts using infiltrants and different types of nano-composite reinforcements.

Curing is a common post-processing procedure for photopolymer materials. During processing, many photopolymers fail to achieve complete polymerization.

As a consequence, these parts are placed in a post-cure system, a tool that floods the part with UV and visible radiation in order to completely cure the surface and sub-surface regions of the part. In addition, the part can undergo a thermal treatment in a low-temperature oven, which can help to fully cure the photopolymer and, in some cases, significantly enhance the mechanical properties of the part.

9.8 PROPERTY ENHANCEMENTS USING THERMAL TECHNIQUES

After AM processing, several pieces are thermally processed to enhance their properties. In the case of DED and PBF techniques for metals, heat treatment is primarily intended to form the desired microstructures and/or to relieve residual stress. Traditional heat treatment recipes developed for the specific metal alloy used are often used in these cases. In some cases, however, special heat treatment methodologies were implemented to maintain the fine-grained microstructure inside the AM component while also providing relief of stress and improvement of ductility. Until the advent of DED and PBF techniques able to directly process metals and ceramics, many techniques for the production of green metal and ceramic parts using AM were developed. These were then post-processed in furnace to achieve dense, usable metal, and ceramic parts production. Binder jetting is the only AM process which is widely for such purpose.

In order to prepare a green part for furnace processing, several preparatory steps are typically done. Figure 9.14 shows the steps for preparing a metal green part made from Laser Form ST-100 for furnace infiltration.

Figure 9.15 shows the injection molding tool made from the ExOne binder jetting process after de-binding, sintering, and infiltration of the furnace. The use of cooling channels following the surface contours (conformal cooling channels) in the injection mold has been shown to greatly improve the injection molding efficiency by reducing

(a) **(b)**

FIGURE 9.14 LaserForm ST-100 green parts: (a) The parts are placed next to the "boats" where the bronze infiltrant is placed. The bronze infiltrates the part through the ships; (b) Parts are often covered in aluminum oxide powder before being placed in a furnace to help support fragile features during de-binding, sintering, and infiltration and to help minimize thermal gradients.

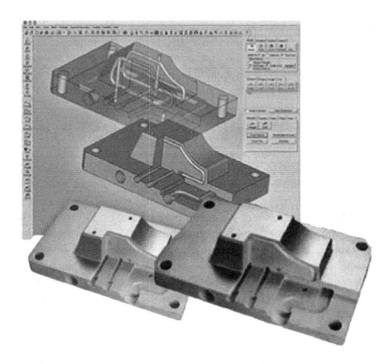

FIGURE 9.15 Cross-section of a ExOne ProMetal injection molding tool showing CAD files and finished, infiltrated component with internal conformal cooling channels.

the cooling time and the part distortion. As a result, using conformal cooling channels effectively allows many businesses to use AM-produced devices to improve their efficiency. Control of shrinkage and dimensional accuracy during processing of the furnace is complicated by the number of process parameters to be optimized and by the various steps involved. Figure 9.16 shows the complicated nature of optimization for this type of furnace processing. The y-axis (F1–F2)/F1 represents the dimensional changes during the final furnace phase of the infiltration of stainless steel (Rapid Steel 2.0) parts using bronze. F1 is the size of the brown part before infiltration, and F2 is the dimension after infiltration. Data represents thousands of measurements across both internal (channel-like) and external (rib-like) features, ranging from 0.3 to 3.0 in. Although many variables have been analyzed, only two have been found to be statistically significant for the phase of infiltration: atmospheric pressure in the furnace and the amount of infiltration. The atmospheric pressure ranged from 10 to 800 Torr. The amount of infiltrant used ranged from a low of 85 percent to a high of 110 percent, where the percentage was based on the estimated amount of material needed to fully fill the porosity of the part, based on the weight and volume measurements of the component just prior to infiltration. This can be seen from Figure 9.16. The factor combinations with the lowest overall shrinkage were not factors with the lowest shrinkage variation. The factor combination A had the lowest mean shrinkage, while the factor combination E had the lowest shrinkage variance. Since the average

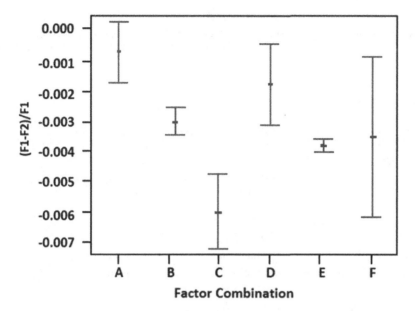

FIGURE 9.16 Ninety-five percent confidence intervals for variation in shrinkage for stainless steel (RapidSteel 2.0) infiltration by bronze (factor combinations are: (a) 10 Torr, 80%; (b) 10 Torr, 95%; (c) 10 Torr, 110%; (d) 800 Torr, 80%; (e) 800 Torr, 95%; (f) 800 Torr, 110%).

shrinkage can be easily balanced by a scaling factor, the optimal factor combination for the highest accuracy and precision would be the factor combination E. It follows that the skin offset d_{offset} would be determined by identifying the shrinkage variation for the whole process (green part manufacturing using AM, plus sintering and infiltration) using a similar approach and then setting the d_{offset} equal to the maximum shrinkage variation at the desired confidence interval.

In addition to the thermal processes discussed earlier, a number of other processes have been developed over the years to combine AM with furnace processing for the production of metal or ceramic parts. For example, laser sintering is used to produce porous parts with gas impermeable skins. By scanning only the outer contours of the part during the SLS manufacturing process, a metal "can" filled with loose powder is produced. These sections are then processed to maximum density using hot isostatic pressing (HIP). This in-situ encapsulation does not result in adverse container-powder interactions (because they are made from the same powder bed), decreased preprocessing time, and fewer post-processing steps compared to traditional HIPs for canned products. The SLS/HIP approach was successfully used to produce complex 3D parts for aerospace applications in Inconel 625 and Ti–6Al–4 V. Laser sintering was also used to produce complex-shaped ZrB2/Cu composite EDM electrodes.

The approach involved:

(a) The production of a green part of the polymer coated ZrB2 powder using laser sintering

(b) The binding and sintering of the ZrB2

(c) Infiltration of ZrB2 sintered porous with liquid copper. This production route was found to result in a more homogeneous structure compared to the hot pressing route.

9.9 EXERCISES

1. Explain natural support post-processing in detail.
2. Explain briefly the synthetic support removal process.
3. Discuss the surface texture improvements process.
4. Describe the accuracy improvements process in detail.
5. Explain briefly the esthetic improvements process.
6. Describe investment casting patterns.
7. Discuss sand casting patterns.
8. Explain other pattern replication methods.
9. Explain other property enhancements using non-thermal techniques.
10. Discuss property enhancements using thermal techniques.

9.10 MULTIPLE-CHOICE QUESTIONS

1. In polymer PBF processes, it is usually important to allow the component to pass through a _____ stage after the component has been constructed.
   a) Heat-up
   b) Cool-down
   c) None of the above
   d) All of the above
   Ans: (b)

2. In natural support post-processing, the part will remain inside the powder to reduce the distortion of the component due to _____.
   a) Non-uniform cooling
   b) Uniform cooling
   c) Non-uniform heating
   d) Uniform heating
   Ans: (a)

3. The cool-down time in natural support post-processing depends on the _____ and the scale of the part(s).
   a) Support material
   b) Filler material
   c) Building material
   d) All of the above
   Ans: (c)

4. When using Quickcast style stereolithography, hollow, truss-filled shell patterns must be drained of liquid _____ to investment casting.

a) Later
b) In-between
c) Prior
d) None of the above
Ans: (c)

5. To seal and reinforce the part for the investment process, the part is pre-invest-ment infiltrated with an investment casting _____.
a) Powder
b) Liquid
c) Granules
d) Wax
Ans: (d)

6. _____ is used to produce porous parts with gas impermeable skins.
a) Laser sintering
b) Binder jetting
c) Drop on demand
d) FDM
Ans: (a)

7. Reproducible shrinkage and distortion can be compensated by _____ the CAD model.
a) Translating
b) Rotating
c) Scaling
d) None of the above
Ans: (c)

8. The AM pattern may be repeated using _____ or electroless deposi-tion processes to create an injection molding core or cavity which can then be used to mold other parts.
a) Spray metal
b) Powder metal
c) Liquid metal
d) None of the above
Ans: (a)

10 Product Quality

10.1 BUILDING THE PART

The start sequence is critical to establish and verify proper process performance before transitioning to fully automated operation. Once it is ensured that all subsystems are functioning together properly and initiated any process monitoring, then process can be initiated. Having an AM operator on call to attend or to a stop the operation is needed to assure proper termination, remove the part, and prepare the system for the next build. Certain AM system designs reduce the downtime between building one component and the next by utilizing dual or modular powder bed configurations, allowing part or powder module removal of a completed build simultaneous with starting the next.

The successful completion of a build cycle often includes a cooling interval particularly for EB, DED, and arc-based systems. PBF-EB, PBF-L, and DED-L are more efficient in melting and experience less heat buildup due to the high-energy heat source. However, PBF-EB uses a powder bed preheat of up to ~700°C and is performed in a vacuum chamber that slows heat transfer and cooling after the build cycle due to the preheating of the powder bed, combined with the additional heat input of the electron beam. An inert gas purge of the chamber may be required to assist in heat transfer and cooling until the part is below a reactive temperature and allowed to cool in air to a temperature that allows effective handling, powder removal, and post-processing.

In the case of PBF-L systems, a shorter cooldown interval is required as powder bed heating is less than with PBF-EB. After cooldown, removal of the powder will include special powder handling procedures to allow removal of the build platform, the part, and support structures.

10.1.1 POWDER RECLAMATION

Powders optimized for AM use are often expensive; therefore, reclaiming and reusing them is critical to the process. The claims that used powder can be re-sieved and reused indefinitely are disputed. Changes to powders may include losses due to re-melting, vaporization, oxidation, and moisture pick resulting from normal processing, or improper handling or storage. These changes may in many cases be insignificant with slight charges to particle size distribution but will also be dependent upon the alloys used and types of powder. This is an active area of research, but now it is best to follow the vendor's recommended procedures. Powder vendors are beginning to fill the need by offering powder characterization services and tracking software to assist customers in maintaining their AM powder inventory.

Removal of the support structures and part from the build plate may require conventional machining, sawing, or EDM (electrode discharge machining) operations as well as finishing the build plate and measurement to assume a minimum build plate thickness and flatness specs are met. Post-build procedures may include cleaning of the chamber or post-processing the powder by sieving to allow reuse and recycle. In some cases, a part may be left connected to the build platform as a support fixture for subsequent finishing operations such as machining or inspection. In other cases, the build plate may become integral to the final part following through all post-process heat treatment and machining operations.

10.2 POST-PROCESSING AND FINISHING

What sorts of post-processing operations (Figure 10.1) are required? Can you perform these operations or will you send the parts to a service provider to complete the post-build finishing operation?

Full functionality of a deposited metal part will most often require post-processing and finishing to achieve the desired dimensions and properties. Post-processing operations for metals may require heat treatments, machining operations, and access to resources that require specialized precision equipment, expert knowledge, and operations beyond simple media blasting, sanding, or coating. Knowledge of these post-processing operations, where and how to access these resources, will help you better choose the process and materials to make the best up-front design decisions. Knowing how and when to apply these operations is critical to efficiently achieve the desired full performance of your metal part.

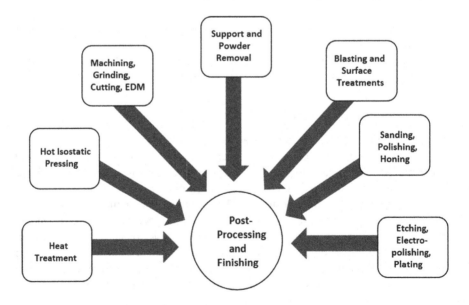

FIGURE 10.1   Post-processing and finishing operation.

One of the first operations after completion of a build cycle and cooling is to remove the part from the powder bed, build platen, or fixtures. As introduced above, this will include powder removal, recovery, and physically removing support structures or fixtures, emptying powder volumes internal to the part through drain holes, and clearing internal passages. Figure 10.2a and b show a support structure and its removal for a nozzle component. Recovering the powder and recycling it back into the process for subsequent builds may require different procedures depending upon the powder type and build conditions. Recommended practices for tracking and mixing new virgin powder with sieved and reused powders will vary depending on applications, standards, and certifications still in development.

Finishing may include media blasting, peening, sanding, abrasive slurry honing, or grinding to smooth surface features and allow visual inspection. Washing may be used to help remove powders from internal features. In cases such as medical applications, sterilization may be specified. Coating and painting, as used in plastic prototype finishing, may improve surface finish or appearance. Alternatively, smoothing or finishing may employ slurry polishing, electro-etch, electro-polishing, or plating operations. These operations may require specialized equipment and processes provided by a dedicated service provider.

Partially fused powder particles may become dislodged and affect in-service part performance. Machining, grinding, polishing, or coating are all candidates to modify as-deposited powder, but the more post-processing required by your design, the more you diverge from having a straight-out-of-the-machine functional object and the more you reduce the benefits of AM fabrication, when compared to conventional processing. Hybrid systems combining improvements in laser optics and beam delivery may extend capabilities to include automated finishing or inspection into the build space. External supports or base features may need to be removed as part of the post-build finishing operations. The design of support structures and optimization to

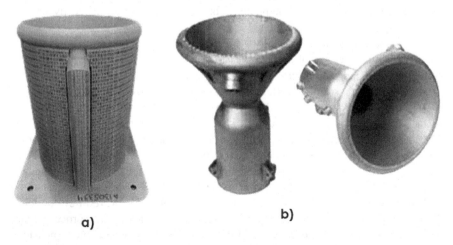

a)   b)

FIGURE 10.2   (a) As-built nozzle structure with support structure; (b) Finished nozzle with supports removed.

allow ease of removal is incorporated into the build cycle using software, while the removal of supports is typically performed by sawing, cutting, machining, grinding, EDM, or other mechanical means. Knowledge of what post-processing methods are best for a specific material and design is critical to optimizing the AM process for a specific part.

Heat treatment is often required to achieve the engineering properties of the AM metal. Heat treatments may require 2 –4 hours in an inert or vacuum furnace at temperatures ranging from 650°C to 1150°C. These treatments may be required to improve or meet the desired strength, hardness and ductility, fatigue, or bulk properties. As stated earlier in the book, heat treatments, such as annealing, homogenization, solutionizing, or recrystallization, may be needed to achieve uniform bulk properties or achieve the desired microstructure. The layer-by-layer deposition can result in directionally dependent properties and can vary with respect to the orientation of the part within the build chamber. HT furnaces used to treat metal parts may need to operate at high temperatures and use inert atmospheres, such as argon or vacuum, when processing certain materials. Relief of residual stresses present in an AM part may also require HT to assure dimensional stability. All these operations require specialized HT equipment and may take hours to complete. Research is ongoing to understand and define heat treatment conditions for AM deposited materials. In one example research indicated differing heat treatments of the same material were required to optimize either hardness or wear resistance.

As mentioned earlier, hot isostatic pressing (HIP) is a process that uses high temperatures and high gas over pressures to heat a part to a temperature below melting and at pressures of 100s of MPa, and temperatures in the range of 900–1000°C for 2 to 4 hours to help close and fuse internal pores, voids, and defects. HIP can also provide heat treatment benefits by optimizing the temperature and pressure cycles to improve mechanical properties such as strength, elongation, ductility, and to improve the structural integrity of the component. The equipment is large and costly and may require a specialty service provider.

HIP pressure chambers typically are limited in size ranging from ~75 mm to 2 meters, limiting the size of an AM part to be consolidated, although custom HIP system designs and services are available in industry.

High precision parts will often require subtractive post-processing operations, such as machining, grinding, or drilling, to achieve the dimensional tolerances and surface finishes of the final functional shape. Specialty operations such as plunge EDM or polishing may be needed to achieve the final surface contours of molds, punches, and dies. The tradeoffs between the accuracy, surface finish, or microstructure of the as-deposited part and the desired final finish of the part entail a complex set of decisions. Regardless of the process used to fabricate the part, either by PBF or DED, laser, electron beam, arc, powder, or wire, a careful evaluation of the final requirements will contribute to the decisions made during process, material, and procedure selection. Creating machining blanks may relax the dimensional requirements of the deposit. In cases such as in DED, deposition rates may be more important than accuracy.

If the as-deposited near net shape requires machining to achieve the final dimensions, surface finish or distortion may be less important than optimizing the build rate. As an example discussed earlier, DED-EB may be used to create very large objects to be subsequently machined to final dimension. The cost savings result from not having to machine very large billets resulting in the creation of a significant amount of wasted material. In cases such as these, the degree of stair stepping, distortion, and relatively crude deposition to the near net shape is less important because the final shape and tolerances will be achieved by machining. The hybrid combination of DED and CNC machining holds promise to incorporate DED using lasers and arc-based systems into multi-axis machining centers. Again, speed may take precedence over accuracy if all the AM deposit will be machined to final dimension.

10.3   BULK DEPOSIT DEFECTS

Figure 10.3 shows some of the common defects present in AM metal parts. Deposit quality is judged by the properties achieved and the ability to meet the design requirements. If you are AM building a component for critical service, such as an aerospace part, part failures can have grave consequences. Therefore, the appropriate level of inspection and quality control becomes important. AM design and deposit requirements are set by engineers with knowledge of what size, number, and type of flaws are allowable and the confidence the process is under control and able to meet these

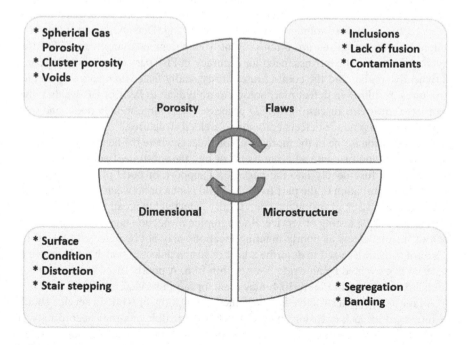

FIGURE 10.3   AM metal flaw and defect types.

requirements. Qualified operators, procedures, formal inspection, and records are as important as the parts themselves.

The functional requirements of components are not limited to structural and esthetic needs. As an example, if high pressure, high temperature, impulse loading, or highly corrosive environments will be encountered by the part, the properties should be appropriate for the service conditions. Knowledge of AM defects that can affect the material properties or lead to failure is critical. Knowing how defects occur and how to detect them, using destructive and nondestructive inspection methods, will help you select the most appropriate inspection methods for these specific types of defects.

Due to the complex nature of the AM processes, the bulk material character or morphology of the deposit can vary widely from one process to the next. It will be different for different materials and may vary widely from one location in the part to another. This may be due in part to changes in the many variables and conditions during deposition. As an example, changes in energy input may change the thermal conditions and the ability to accurately deposit certain features. Defects not within the bulk material, but within the character of the part, such as distortion, may also occur.

Flaws are undesirable conditions within the deposit, while defects are flaws that exceed acceptance criteria. The detection of defects can trigger actions such as rework, repair, or part rejection. AM metal defects may include lack of fusion, porosity, voids, cracking, oxidation, discoloration, distortion, irregular surface profile, or surface stair stepping. Later we discuss common AM inspection techniques including visual inspection, dimensional inspection, dye penetrate, leak testing, radiographic, computed tomography, and proof testing. Defects typical to fused or sintered metal processing are often present to one degree or another in 3D fused metal parts. Given the intrinsic need for accuracy in PBF parts, the molten pool size is typically smaller and the cooling rates are typically faster than common welded structures, resulting in defect morphology more typical to laser or EB welds rather than large arc welds or castings. DED processes with large molten pools and high deposition rates produce defects common to welded structures.

Lack of fusion is one of the more common defects where the bead or deposit does not fully melt and fuse into adjacent tracks or into the substrate. Lack of penetration is a failure to fuse deeply into the part, build support, or build platen, which may result in de-lamination of the part from the build platen or between layers within the deposited part. Lack of fusion may be visually detected but it can also go unnoticed until a part fails in testing or service. Radiographic inspection may or may not detect lack of fusion defects as poorly bonding locations may not be detected. A cold lap is a term sometimes used to describe a lack of fusion defect in which coalescence of material is prevented by an oxide layer or thin film. A poorly fused powder particle boundary, as shown in Figure 10.4, may crack or fail in service.

A failure during destructive testing, functional testing, or while in service should be inspected for lack of fusion and the AM build schedule modified accordingly. In some cases, flaws leading to fracture initiation points may only be observed under high magnification using scanning electron microscopy (SEM) and a metallurgical

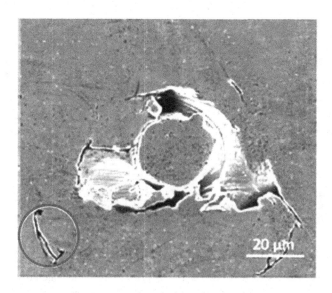

FIGURE 10.4   Lack of fusion defect showing unfused powder particle.

inspection technique known as fractography. We will discuss these later under failure analysis.

Slumping is a dimensional defect typically created in regions built with insufficient support structure. Slumping defects may be associated with small design features, thin walls, downward facing surfaces (also referred to as down skin surfaces), or overhangs. If the molten pool is too large for the position, gravity can sag or slump the molten pool and create this type of defect. Rounding or loss of edge quality may occur when surface tension draws a molten pool into a rounded shape resulting in the loss of dimensional fidelity of sharp edges or corners. If the heat source power is too high and deposit or traverse speed is too fast, surface tension can draw the weld bead up into a ropey shape leaving a lack of fusion along either side of the deposit. These sorts of instabilities may also lead to spatter and droplets of melt or partially melted powder particles from the melt region, further disrupting the deposition quality. Proper parameter selection such as control of beam power or speed can help to fuse and flatten the deposit bead. Undercutting is a flaw typical to fusion processing that can occur in regions where gravity or surface tension draws molten metal away from the edge of the pool surface leaving an underfilled region that may act as a stress riser or crack initiation point.

Shrinkage and distortion can occur as a result of localized melting and solidification. Shrinkage-induced stresses can build up creating warping and distortion and in some cases reach levels where lack of fusion and de-lamination can result as shown in Figure 10.5. Proper design, material and parameter selection, and process development can reduce or eliminate the risk of these types of defects.

Irregular surface condition can indicate a poorly designed, supported, or oriented part. A poorly developed or controlled process may result in stair stepping, balling, and lack of fusion, surface breaking de-lamination, undercutting, holes, porosity,

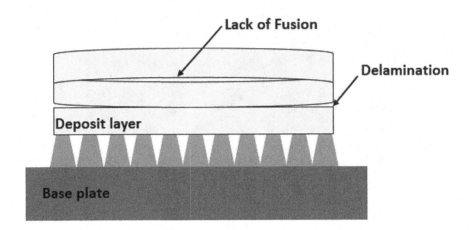

FIGURE 10.5   Lack of fusion or delaminating defects.

or voids. Excessive spatter of balls fused powder particle ejected from the melt region may be fused to either side of the deposit, indicating improper beam power, focal conditions, contaminated powder, or filler wire. Irregular surface condition can reduce part strength by creating stress concentration locations that may fail or fatigue during service. Notches, voids, or undercuts can concentrate stresses and initiate cracks. Cosmetic and surface requirements not met will require post-processing to either remove or finish the top surface.

Porosity is a common defect evolved in melted and fused material, often spherical or oblong, resulting from gases, such as hydrogen entrapped within the molten pool and evolved and released as bubbles upon cooling and solidification. Another source of porosity is the melting of an un-fused region or void in which the gas within the void forms a bubble within the liquid and entrapped within the solidified metal. Keyhole collapse as in laser or EB melting can entrap gas within the melt pool during solidification.

It is important to differentiate between gas porosity and other forms of voids such as lack of fusion voids as different mechanisms lead to their formation and need to be considered separately to assure proper control. Sources of these gases may be moisture or contamination of the inert gas atmosphere, build chamber, powder, or filler supply. Hydrogen as a gas is readily absorbed into molten metal and can be rejected from the melt during solidification and trapped as bubbles or pores. Hydrogen contamination of the feed stock or the process before or during the build cycle can result from a number of sources such as improper quality control of the wire drawing process or improper storage. Porosity is more of a problem for some materials than others, such as aluminum and those more susceptible to gas absorption during powder or melt processing. Strict handling, storage, and processing procedures are needed to control sources of porosity. Argon may be trapped into powder as a result of the gas atomization process and may be another source of micro-porosity. Entrapped gases such as these may coalesce during melting and result in the growth of porosity during solidification.

Pores can reduce the cross-sectional area of the deposit, thereby reducing strength, although spherical porosity is not as critical to loss of strength as angular crack initiating defects that have a greater tendency to propagate under loads. Surface breaking porosity and voids can hold water or moisture and exacerbate corrosion and staining. The AM designer or fabricator needs to consider all possible defect formation scenarios to assure that the requirements of the deposit are met and to decide what level of porosity or flaw content is acceptable. The distribution of porosity within a fused region may also help to identify its origin. Spherical porosity may indicate a contamination source either within the powder, atmosphere, or resulting from contamination during storage, handling, or processing.

Figure 10.6 shows two levels of porosity in an SLM deposit of AlSi10 Mg achieved by drying the powder and modifying the process parameters.

Voids not evolved from gas rejection or gas entrapment upon solidification may take many different forms and can be the result of many different processing conditions. Lack of fusion voids common to the AM PBF processes include those associated with the powder packing density, or the spaces in-between powder particles, when spread in layers and inadequately fused, resulting in a poorly fused deposit. The microstructure of sintered powders, resulting in less than fully dense deposits, will display un-fused regions or voids requiring re-melting through subsequent layers of deposit or additional HIP post-processing. These regions may contain un-fused powder particles or inadequately fused particles due to insufficient liquid phases while sintering, or in other cases insufficient melting or mixing within the molten pool. As an example, a decrease in beam energy density may result from an optical component that is misaligned or needs cleaning, resulting in a lack of fusion defect. Conversely, increased energy density resulting in a deviation of scan speed, e.g. during a change in scan direction, may create a localized vaporization event or an unplanned transition to a keyhole mode of melting resulting in entrapped porosity due to keyhole collapse.

Localized inclusion defects may result from contamination contained within the powder or filler wire which may vaporize during processing with sufficient force to eject molten material from the melt pool leaving a hole or void that may not fill during the fusion of subsequent layers. As an example, erosion or damage to the

 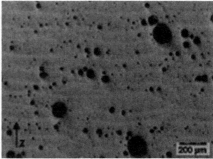

FIGURE 10.6   Hydrogen porosity within SLM deposit of AlSi10Mg.

recoating blade powder spreader or rake may leave particles of foreign material within the build material supply. Constraint is a term often used when discussing the formation and avoidance of cracks associated with metal fusion. As discussed earlier in the book, metal expands or contracts when heated or cooled. As a result it may grow, shrink, warp, or bend.

If heated or cooled and constrained from moving, such as by clamping, the use of support structures, or because mechanical constraints within the design itself, stresses will build up. These stresses are mechanical forces that can either be locked up and reside in distorted crystalline structure, or be relieved by distortion, cracking, or tearing. The degree to which movement is allowed or prevented can be referred to as constraint. Residual stresses are difficult to measure and although not often classified as flaws, may lead to distortion or cracking in service. They are often controlled through the part design, processing condition selection, or post-build heat treatments. Cracking can result from a wide range of thermal, mechanical, and metallurgical conditions. Figure 10.7 shows cracks in a welded reactor vessel as detected by dye penetrant testing. AM components can be susceptible to the same types of cracking mechanisms present in welded or weld clad structures. A few of the more common types of cracks are crater cracks, hot cracks, hot tearing, and cold cracking. Some materials are much more sensitive than others to the development of cracks as a result of AM metal processing. Crater cracking refers to cracking that may occur at the termination of an AM deposition path within the last to solidify material. Hot tearing may occur in regions directly adjacent to the fusion boundary, at temperatures below melting, in the metal softened by heating. Hot cracks may occur directly upon cooling near the solidification boundary, while cold cracks (also known as delayed cracking) may occur hours or days after cooling.

The metallurgical reasons for these types of defects range widely and are beyond the scope of this book, but the reader should know these conditions exist and consult

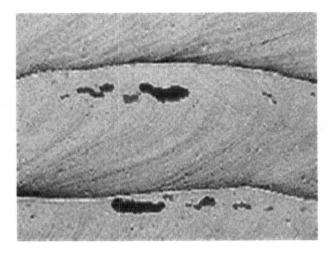

FIGURE 10.7   Cracks in a welded reactor vessel as detected by dye penetrant.

a professional metallurgist when working with new alloys or with potentially crack sensitive materials. AM vendors have developed strict control of materials and parameter sets to avoid many of these types of problems, but if you are developing your own parameters for materials that are crack sensitive, be ready to address these problems. Cracks can occur at the termination or end point of an AM melt track or weld. The last part of the bead to solidify results in a depression and shrinkage and may result in cracking. A crater depression with cracking can concentrate stress induced by in-service loads and propagate the cracks possibly leading to failure. Properly developed procedures, such as those supplied by vendors, and proper selection, and control of materials will avoid these types of defects.

Hot cracks and tears form as the molten pool solidifies. They are often formed by shrinkage stresses induced during cooling due to a combination of chemical, metallurgical, and mechanical conditions of the solidification region, partially melted or low ductility heat affected region. Highly constrained part locations, unable to flex and distort due to thermal expansion and contraction, can literally pull apart or tear when the metal is still hot and weak during cooling. Recall the discussion early in the book regarding thermal softening of metals. As with porosity, crack location and character can indicate the mechanism of formation and assist in detection, prevention, and control. As with fusion welding, many types of cracks may be associated with AM deposits. Figure 10.8a,b shows micro-cracking in direct metal laser sintered Inconel 718 produced under experimental conditions. Micrometer scale cracks can occur along grain boundaries and across grains and backfill with the last to solidify material making them hard to detect. Proper parameter selection, preheating conditions, or HIP processing can reduce or eliminate the formation of cracks and bond defects in crack sensitive materials.

Undetected micro cracks within components under service conditions, such as cyclical loading, may eventually lead to crack propagation and in-service failure.

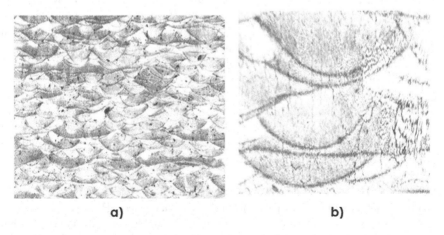

a)　　　　　　　　　　　　　　　　　b)

FIGURE 10.8 (a) Microcracking of DMLS Inconel 718; (b) Microcrack opening in Inconel 625 shown using scanning electron microscopy.

Hot cracking is commonly encountered in aluminum alloys but can often be avoided by proper selection of part geometry, metal, and parameter selection. Stress relief features may be designed into the part or support structure to relieve stresses near crack susceptible regions. Cold cracking, as the name implies, refers to cracks that can form hours, days, or much later in the service life of a component with the potential for catastrophic failure. There are many more types and reasons for cracks, but suffice it to say that cracking in AM deposits may be one of the more serious defects, as a small crack initiation site can result in catastrophic failure in service. Earlier we discussed some design techniques used to avoid distortion and cracking. Later in the book, we discuss some ways to detect cracks and avoid cracking in AM parts as the recommended practices being developed may deviate from those of conventional processing, such as those typically used for welded fabrication.

Oxidation and discoloration can occur when the molten pool or surrounding hot metal is improperly shielded from air and atmospheric conditions. Chemical reactions take place at different temperatures forming compounds of different colors on the part surface. Changes in chemistry may also occur below the surface. Changes in the build schedule, process disturbances, interruptions, or other sources of contamination can affect the surface chemistry. Regular maintenance, source material control, and proper setup may help reduce sources of contamination and discoloration. Routine use of equipment checks to look for problems, such as loose fittings or problems with the gas supply, is also warranted.

10.4 DIMENSIONAL ACCURACY, SHRINKAGE, AND DISTORTION

The accuracy of a desired part is largely determined by its design, processing conditions, and material. As mentioned earlier, the designer must take into account the thermal and mechanical characteristics of the material being processed. The build sequence or schedule must appropriately specify deposition conditions to account for localized thermal expansion and contraction during the build cycle. Contouring conditions, or those build parameters used along the perimeter of each AM layer, must be tailored to achieve the desired accuracy and surface smoothness. Orientation of the part within the build chamber, to minimize stair stepping and layer effects, is needed as well as the appropriate design of support structures used to assist in the deposition of overhang features. Spring back of a part when removed from a base platen or support structures may need to be accounted for in the design.

When building a near net shaped component that will require subsequent machining, a material or machining allowance is required. As an example, the material allowance enables the machining process to remove enough of the as-deposited surface to reach sound metal and achieve the required surface condition. For parts requiring high-dimensional accuracy, a machined datum or inspection surface is required to define the location of other inspection features. Another way to describe this is "being able to find the part within the near net shape." Heat treatments such as stress relief, annealing, or HIP processing may induce additional dimensional changes which may require revising the original design, as dimensional changes resulting from heat treatments may be unpredictable during the first design iteration.

In the future, computer simulation and prediction tools, such as FEA modeling, may be developed to assist in accurate shrinkage prediction for large complex parts. But for now, trial, error, and experience may be the only option. When it comes to bulk material or part defects, knowing what you are looking for and how defects are formed is key to detection and prevention.

10.5 INSPECTION, QUALITY, AND TESTING OF AM METAL PARTS

Additive manufacturing processes used for metals present some unique issues associated with inspection, quality assurance, and testing. As any metal fabricator knows, detecting flaws and defects in a weld is difficult enough, but when the part is entirely made up of welded filler material the potential for those types of defects may be present in any location within the part. The refined nature of the buildup, due to thin deposition layers and fast cooling rates, will lead to defect morphology and geometric conditions that challenge even the best inspection methods currently in use. In some cases, current inspection methods applied to AM produced parts that are fully heat treated or HIP processed and 100% machined may be appropriate, in other cases alternate means for acceptance may need to be developed. As a general rule, as part complexity increases NDE inspect ability decreases. AM processes can add significant design complexity challenging traditional NDE techniques. In this section, we discuss available inspection methods and some of the challenges presented by AM metal fabrication.

10.5.1 Nondestructive Test Methods

Figure 10.9 identifies nondestructive test methods being applied to AM parts. Visual inspection is often employed to identify gross defects such as distortions, surface conditions, and gross anomalies. Camera and image inspection can accurately and rapidly capture and classify part features and compare them with a standard definition or part model. Cameras may also utilize magnified views of regions to verify conditions against a standard or population of parts. Bore scope inspection may be incorporated to verify clearances or deposit conditions within enclosed volumes or passageways. Multiple image cameras combined with software are capable of measuring distances and other characteristics of AM parts.

In another example of visual inspection, titanium surface discoloration can indicate contamination of the powder before or during processing and may be cause for part rejection. Supporting this example, the AWS, Specification for Welding Procedure and Performance Qualification, specifies color acceptance criteria for class A, B, C critical welds. Challenges to visual inspection presented by AM can include complex shapes and internal feature beyond the line of sight of inspectors or cameras. Unless the part was specified to be built in the same orientation, using the same support structures and build parameters, the occurrence of surface defects, such as stair stepping, may occur in different locations on parts built to the same 3D model.

Dimensional inspection relies on measurement tools for dimensional verification, while gauges may offer a faster way to verify a part dimension or extent by

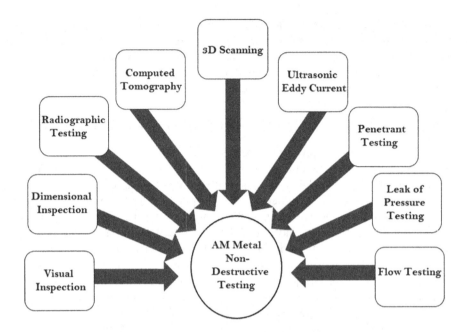

FIGURE 10.9 Nondestructive testing techniques for AM.

using a GO-NO-GO gauge. Various nondestructive methods may be used to assure the clearance of internal passageways and volumes such as by using displaced liquid, pressure–volume–temperature, flow rate. Coordinate measurement machines (CMM) are precision measurement devices employing a touch probe, manipulated using CNC or robotic motion, to measure the location of part features.

Software may take these measurements and compare them to locations on the part definition model. On-machine gauging can incorporate an inspection gauge directly into a CNC milling or turning environment. With the advent of hybrid AM/SM machines, the logical next step is combining CNC with AM and in-process measurements. Methods such as this are currently under development potentially enabling a single machine to produce a fully certified part without the need for subsequent inspection steps.

3D digital scanning refers to a number of techniques used to capture the geometric extent of an object, generate a point cloud of data approximating the surface of an object, and fit a geometric surface or solid description to the point cloud creating a digital model. This point cloud generated solid model may then be compared to the original part model definition. Digital scanners often are structured light or laser-based systems that move in relation to the part surface sensing the reflected beam to define the location of data points in 3D space. Surface finish, color, reflectance, smoothness, and lighting can affect the accuracy of the scan and may limit the effectiveness in certain AM applications. As with visual inspection, complex shapes with no line of sight to critical features may eliminate the usefulness in certain AM applications. Deep features, high aspect ratio holes, and internal features may not be captured. Commercial scan systems range widely in accuracy and costs as does the supporting software. Software is used to best-fit the point cloud to geometric features and surfaces, but again, the accuracy of the point cloud and the fitting algorithm will affect the final definition of the inspected part. Radiographic testing (RT) has been used for many years to inspect welds in pipelines, pressure vessels, and a wide range of critical use metal components.

Irregular surface conditions or multiple or complex internal features along the X-ray path may obscure features of interest or complicate interpretation of the images. Digital radiography offers a 2D grayscale image that may be enhanced using color or other digital techniques. Micro-focus radiography offers greater resolution for a given wall thickness within a narrow field of view. Computed tomography (CT) relies on a series of images, such as those obtained by X-ray or magnetic resonance imaging (MRI), taken at specific angles relative to the part and reconstructed by computer software into a 3D data set revealing interior and exterior features. CT inspection models may then be used to compare to the original CAD model enabling part-to-part comparisons (Figure 10.10). The quality and resolution of the scan and the accuracy of the algorithms used to reconstruct and render the geometry will affect the results. The energy source, detector resolution, and setup will contribute to the measurement accuracy. These methods have been demonstrated to characterize features such as porosity, inclusions, minimum wall thickness, and other internal features. One problem can be that of false positive indications of defects due to the complex microstructure of certain as-built AM parts. While this technology has

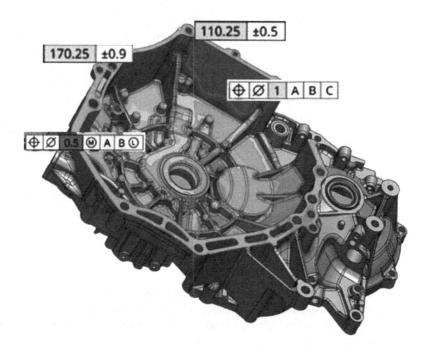

FIGURE 10.10   CT CAD to inspection model using Geomagic Control X software.

been successfully demonstrated in the analysis and inspection of AM metal parts, the costs, complexity, and computing requirements of the large data sets generally restrict the use to specialty service providers and for use by customers with high value components.

Ultrasonic testing (UT) uses ultrasonic sound waves to penetrate a metal object, reflect off internal features, and bounce back to a detector to reveal the approximate size and location of these features. It generally relies on a liquid coupling between the probe and the top surface of the object. It may be limited by curved, complex internal and external surfaces, or the rough surfaces of AM deposited parts. Post-processing and finishing may be needed prior to the application of UT. Penetrant testing (PT) is often used to detect cracks or small surface breaking flaws. It uses a liquid penetrant, often a dye, sprayed on the surface to be tested, that is absorbed into the flaw features. The excess surface dye is wiped off and a liquid developer coating is sprayed on to dry and absorb penetrant from within the cracks, crevices, voids, or pores revealing the defect location (refer to Figure 10.7). While the method may be limited for use on as-deposited AM surfaces of powder-based systems, due to surface roughness, it can offer an effective method to detect cracks, lack of fusion, or undercuts in weld deposits of DED wire-based processes. Magnaflux testing is often used on large cast components to detect cracks in steel and other magnetic materials. It is often used to find cracks in engine components but may find application as applied to AM repaired parts.

Eddy-current testing (ET) is used to detect surface and subsurface flaws, such as cracks and pits, in conductive materials using electromagnetic induction. It can be sensitive to surface and near surface conditions and material type. Vacuum leak testing can be an effective way to assure the hermetic seal of a volume or containment vessel. Complications of applying this technique to AM fabricated products include the need for a sealing surface or sealing compound with the rough surface of as-deposited AM parts or the post-machining needed to obtain a flat sealing surface. Sintered, porous product, or partially fused powder surfaces could create a virtual leak or hidden pumping volume, reducing the utility of the process.

Flow testing may offer a means to assure clearance of complex passageways without the cost and complexity of a CT system. While many of these NDT procedures may be difficult to apply to as-deposited AM parts, standard test specimens and standardized acceptance criteria will be developed that offer the ability to characterize the bulk deposit using NDT methods combined with DT (destructive test) methods, as described below. A good source for additional information is ASNT, the American Society for Nondestructive Testing.

10.5.2 Destructive Test Methods

Earlier in the book, within the process development section, destructive testing was mentioned as being used during parameter studies and during the fabrication of test samples. The samples are sectioned and subjected to a variety of metallographic tests to determine the micro-structural character, defect morphology, and mechanical properties of the deposit as they relate to processing conditions. Small lot production may rely on destructive testing of witness coupons fabricated layer by layer or within the same build cycle of the actual part to infer proper machine function. Figure 10.11 lists categories of destructive test methods being applied to AM parts.

Micro-structural analysis will often entail preparing small samples of representative regions of interest within a part and by mounting, polishing, and in some cases etching the samples to reveal the structure of grains, phases, inclusions, and defects such as pores, voids, cracks, and other discontinuities. As-polished samples may rely on camera-based microscopic inspection, coupled with image processing, to provide deposit density estimation, (% voids), or other semi-quantitative determination.

Hardness values and phase identification can be used to verify time–temperature transformations related to cooling rates and infer property determination confirmed by standard mechanical test procedures. Micro-structural analysis coupled with non-destructive testing and proof testing is often all that is needed to qualify a noncritical component or process. A source for procedures used for micro-structural analysis can be found in the tech sheets of the Beuhler.

For common engineering materials in commercial shapes, mechanical property data contained within engineering data books such as those provided by the Society of Automotive Engineers (SAE), military specifications (MIL), and the American Society of Mechanical Engineers (ASME) is often sufficient to provide the confidence and documentation required by a quality trail. In cases where standard test data is unavailable, a full material and process qualification may need to be performed to assure the material properties and part performance. Vendors of AM

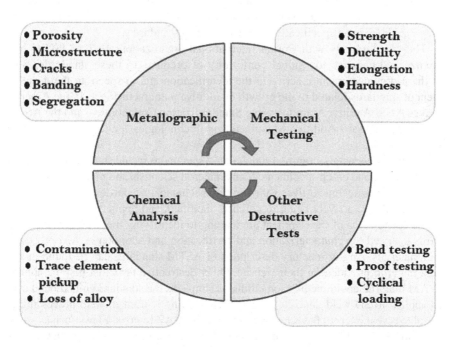

FIGURE 10.11   AM metal destructive tests.

machines often supply this information for their proprietary material, produced using their standard parameter sets and using their stated test conditions. In some cases, this may be sufficient to infer the integrity of a functional prototype, although the material test conditions and properties cited may not be representative of those of any AM produced part.

Service providers of metallurgical and micro-structural analysis are readily available via the Web and can provide rapid turnaround for a full range of services to characterize and formally perform analysis for process development and qualification needs. Design definition of standard test specimens may be enhanced to include standard AM deposition and testing conditions with which to compare material lot chemistry variations or machine-to-machine differences.

Chemical analysis can be applied to verify the integrity and purity of materials and processing conditions used in AM. Samples may be sent to service providers to identify levels of contaminants such as oxygen, hydrogen, nitrogen, or trace elements such as iron, vanadium, or aluminum. This analysis can be useful to determine the pickup of contaminants or loss of alloying constituents. Results may be compared with feed powder or wire chemistry specifications as required for process or part acceptance. Standardized mechanical test specimens and testing procedures have been used for technical generations to assist in alloy development and process material and process certification. ASTM International (referred to as ASTM) leads the development and delivery of voluntary consensus standards to improve product quality, enhance health and safety, strengthen market access and trade, and build consumer confidence. The American National Standards Institute (ANSI) serves to administer and coordinate the private-sector voluntary standardization system, for creation, dissemination, and use of standards and specifications for United States' industry.

They work closely with similar international organizations such as ISO to harmonize and enhance the global conformity of products to these standards. Both of these organizations are active in the identification and to speed up the development of standards related to the growth of additive manufacturing as in the America Makes ANSI Additive Manufacturing Standardization Collaborative and the ASTM Committee F42 on Additive Manufacturing Technologies for the development of AM standards.

As applied to metals, common standard test specimens include tensile bars, Charpy V-notch tests, and creep specimens. Many material scientists and materials engineers have spent their technical lives studying the mechanical properties of metals. These test methods are actively being researched, modified, and applied to AM materials. These active areas of development are leading to identifying the need for new standards to speed the characterization and certification and adoption of AM processes and materials. An introductory description of ASTM standards can be found in student material provided on their website. Other destructive tests that may be applied to AM material could include something as simple as the standard guided bend tests as applied to 3D weld deposits. Guided bend testing is used in weld processing or metal to reveal internal flaws in weld deposits and HAZ regions. Flaws such as voids, cold laps, lack of fusion, poor fusion, porosity, or other micro-structural defects can be revealed in surface or near surface regions. Flat plate samples taken from bulk

AM coupons may be surface machined and bend tested representative of orientations related to the X, Y, Z orientation of the part within the AM build chamber.

Nondestructive testing of machined test coupons such as those described above may lend themselves more readily to inspection by UT, RT, and PT. While not testing an actual part, these samples can infer a represented characterization of bulk AM material, parameters, and procedures as qualified for an actual build procedure. As you can see from the discussion above, the inspection of AM parts remains a challenge as the procedures and technology are still being developed.

10.5.3 Form, Fit, Function, and Proof Testing

Form and fit of prototype designs was the original function of 3D-printed parts. The ability to visualize and handle a 3D part, rather than crowd around a set of blueprints or a computer-generated model, has advantages, especially for customers and stakeholders not intimately associated with the technical design phase. The same goes for fitting a 3D model into a prototype system. While 3D CAD models can be assembled into virtual systems to simulate part and system function, there is great value in someone actually bolting up a part and confirming they can swing a wrench at it. The prototype does not need to be metal to add value to the form and fit stage of prototype development.

Composite materials are blurring the boundaries of plastic and metal used in prototypes for fit testing and functional testing. Advances in materials offering high temperature and strength, such as metal filled plastic composites, have been applied to offer an increasingly wide range of functional testing. The benefit of these hybrid materials is primarily their lower cost. Metal prototypes go one better by potentially offering the testing of the full functional range of part performance albeit at a greater cost. Metal prototype parts may also be used in short run production such as in molds, dies, tooling, and inserts used to produce other prototypical parts by conventional means such as injection molding or casting without incurring the high cost of large production run tooling.

Full functional testing of rapidly produced prototypes, saving both time and money, has been the primary function of AM technology. Although many applications are moving into full production, prototyping will still demonstrate the technology. Large corporations with highly skilled technical teams are routinely using this technology for larger and larger parts in an increasingly wider range of materials. The boundaries of what the technology can do are being pushed at a very rapid rate. Smaller businesses are finding they need to rely on service providers to tap into this market to stay competitive. Applications are emerging where full production of small lot size components is possible, but for now, keeping up with the competitor's rush to market by using AM is sufficient motivation for small businesses to explore the use of AM. Due to cost and limited availability of engineering skills, AM machines, and specialized facilities, small- and medium-sized businesses are turning to established service providers to meet their needs. In some cases, large corporations are acquiring or merging with these service providers to establish in-house skills or to deny access to these services by their competition.

Proof testing is a way to subject a prototype part, a process qualification lot, or a random sample taken from a production lot, to a mockup testing of the part within an in-service environment. As an example, this is the testing stage where you fire up the rocket engine nozzle, take the car to the race track, or pressurize the storage vessel to see how well it works. Often the component is subjected to an extended range of standard operating conditions to prove functional performance beyond that typically seen in service. Elaborate machines to perform these tests have been used for years to evaluate test parts destined for manufacture. We have all seen the so-called shake, rattle, and roll systems where you bolt up the car and subject it to high cycle fatigue for tens of thousands of loading cycles and we all know the fate of the crash test dummies. It will probably not be too long before these machines and dummies feature 3D-printed parts.

10.6   STANDARDS AND CERTIFICATION

Prototype components used for in-house testing and evaluation are often built with a lesser degree of formality and quality documentation that is needed for certification of production parts used in critical applications. Risks associated with the use of prototype AM hardware built for in-house use can be better managed and controlled than those sent to a third party or customer for testing or use. Prototypes are often fabricated without formal operator training, specifications, standards, or certified materials. Costly inspection techniques may not be used at this stage of development.

However, production items sold to the public or introduced into the commercial market need the additional rigor and confidence required by both the producer and consumer. Critical components, or systems of components, such as flight critical hardware used for aerospace, automotive components, energy, or medical applications all need to adhere to common standards, and meet regulations or codes set by governing bodies. Organizations such as the American National Standards Institute (ANSI) promote, facilitate, and approve the development of standards, while organizations such as ASTM International (ASTM), International Organization for Standardization (ISO), the American Welding Society (AWS), and the Society of Automotive Engineers (SAE) assemble committees of experts to lead the development of voluntary consensus-based standards. Governing bodies that license and certify persons, organizations, products, and services include the Federal Aviation Administration (FAA), the Food and Drug Administration (FDA), and the Occupational Health and Safety Organization (OSHA) to name a few. The reader is encouraged to search the weblinks of these organizations to learn the latest AM related activities of these organizations. Regulations, rules, standards, and recommended practices have evolved over decades to apply controls to the production of hardware, feedstock materials, and the qualification of practices and procedures for the benefit of industry, individuals, and the public at large. Certification of metal components used in critical applications and systems is a long and evolved process, with cradle to grave controls and documentation, for materials, qualification of fabrication processes, procedures and operators, and archive of inspection and quality records. As an example, the design of a new turbine blade may cost in the order of $10M, and certification of the new blade may cost \$50M, while the certification of a new system such as an aircraft engine could cost \$1B. Figure 10.12 shows a "System V" development cycle for the verification and validation of components and systems. This type of development cycle can be modified for a wide range of components and systems with the ultimate goal that maps the process of how verification is used to assure parts meet the specified design requirements and how validation is used to assure the part functions as intended.

All of these organizations are actively involved in developing and modifying certification procedures and processes to accommodate the use of AM fabricated parts for a wide range of new materials and processes. Solving the AM metal part certification challenges of today is many ways lagging behind the pace of technology development, demonstration, and the increasing rate of market acceptance.

The economics and cost benefits of moving to the production of certified AM parts are being pulled by demand. There remains, however, a large gap between the production of test coupons, functional test hardware, and fully certified parts. An increasing number of AM parts for use in the medical and aerospace fields are being certified for use on a case-by-case basis. This one-by-one process is costly and time consuming but provides increased justification to identify new industry standards needed to apply AM fabrication to a wider range of components.

For those not familiar with production requirements, here are a few. Some of the formalities of operations needed for the move from prototyping to production include equipment calibration, operator certification and development of standards, ISO 9000 quality control, documentation, record-keeping and formal specifications, working to standards, and qualified materials operators and suppliers. For those not aware of these requirements, they can easily exceed the cost of design, prototyping,

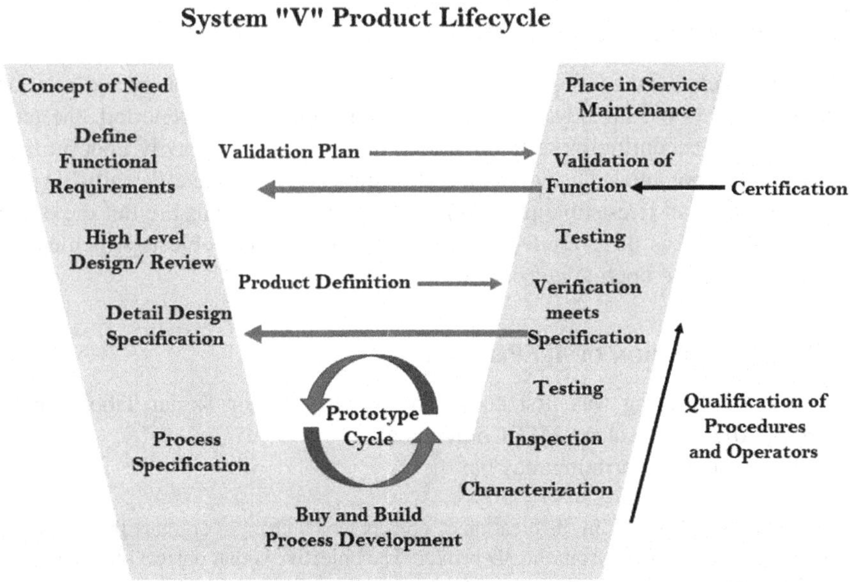

System "V" Product Lifecycle

FIGURE 10.12   System V product life cycle.

and process development. The costs of launching a product into production can be huge and if the production volumes are low the price per unit must be very high. The certification process may also require additional component builds, and verification that parts meet all requirements, regulations, and specifications, such as those required for testing and process verification followed by functional validation to assure customer requirements are met. Certification for AM may require automated systems specifically designed and built to automatically characterize AM materials to populate material property databases.

10.7 KEY TAKEAWAY POINTS

i. The sequence of operations associated with AM metal processing is differentiated from other metal processing operations that require manual intervention or constant supervision. Many AM metal processes can take hours to build a part but require expert oversight to assure the process is properly set up, the safety envelope is verified, and the procedure is followed.

ii. Post-processing operations, inspection, and quality assurance rely primarily on the modification of existing inspection methods to assure a repeatable process, allowable levels of flaws, and a component that meets quality standards. The size, distribution, nature, and origin of AM metal defects in some cases challenge existing NDE methods.

iii. Standards are being developed to assist in the certification of critical components and are needed to fully realize the adoption of AM metal processing in wide-scale industrial production.

10.8 OVERVIEW OF 4D PRINTING

Additive manufacturing (3D printing) is itself an emerging technology and is in fact over thirty years old already. As SPI Lasers has continuously reported, the technology is now becoming more mainstream, but is still very heavily underutilized considering its potential. The potential to economically and time efficiently 3D print ANYTHING is an irresistible proposition. 3D-printed materials are not the end of the story though, as there are techniques to create materials/objects which can be pre-programmed to operate in a certain way.

10.8.1 A DEFINITION OF 4D PRINTING

The term 4D printing was first coined by TED professor Skylar Tibbits in his February 2013 speech at the MIT Conference.

A definition of 4D printing may be:

> The use of a 3D printer in the creation of objects which changes or alters their shape when they are removed from the 3D printer. The objective is that objects are made to self-assemble when exposed to air, heat, or water; this is caused by a chemical reaction due to the materials utilized in the manufacturing process.

10.8.1.1   The Difference between 4D and 3D Printing

Think of 4D printing as the same as 3D printing with the addition of time. By adding time to 3D printing, the concept of 4D printing is born. This enables objects to be pre-programmed in various ways to react to a range of different stimuli.

4D printing is futuristic but has a very exciting future. 4D printing delivers the possibility of designing any transformable shape, which can be made from a large selection of materials. These different materials will have many different properties and a range of potential applications and uses. There is a real opportunity for the creation of dynamic self-assembling objects which could transform and be used in a wide range of industries and in a large number of applications.

10.8.2   POTENTIAL APPLICATIONS FOR 4D PRINTING

Applications of 4D printing are particularly suited to changes in environmental circumstances:

Architecture: Buildings which are delivered in a flat pack form but entirely self-assemble when the right stimuli are added.

Clothing: Clothes and footwear which change the appearance and function (e.g. clothes which naturally adapt/change to the size/contours of the wearer).

a)   An example is shoes which become waterproof during rain or react to other external atmospheric conditions

b)   Military clothing, e.g. clothes which camouflage, cool, and/or insulate soldiers by reacting to different input environments

Food: The 4D printing of food using a number of techniques.

Health: Multiple applications including nanotechnology uses:

a)   There is the possibility of inserting implants into the human body, which self-deform to a plan when inserted with surgical intervention (e.g. cardiac tubes)

b)   Using 3D printers injected with stem cells to print slices of liver and other organs

c)   Using 3D printers to print skin, the shape of which changes over time depending on conditions

Home appliances: Products in the home, such as a chair which upon purchase self-assembles through heat stimuli applied by a home hairdryer.

Transport: Roads which self-heal potholes.

10.9   EXERCISES

1. Briefly explain post-processing and finishing.
2. Discuss nondestructive test methods.
3. Discuss destructive test methods.
4. Explain standards and certification.

10.10 MULTIPLE-CHOICE QUESTIONS

1. PBF-EB uses a powder bed preheat of up to _____.
   a) 600°C
   b) 700°C
   c) 800°C
   d) 900°C
   Ans: (b)

2. HT furnaces used to treat metal parts may need to operate at _____ and use inert atmospheres, such as argon or vacuum, when processing certain materials.
   a) High temperatures
   b) Moderate temperatures
   c) Low temperatures
   d) None of the above
   Ans: (a)

3. HIP is a process that uses high temperatures and high gas over pressures to heat a part to a temperature below melting and at pressures of _____.
   a) 80 MPa
   b) 90 MPa
   c) 100 MPa
   d) 110 MPa
   Ans: (c)

4. _____ inspection may or may not detect lack of fusion defects as poorly bonding locations may not be detected.
   a) Radiographic
   b) Heat treatment
   c) Fractography
   d) All of the above
   Ans: (a)

5. Ultrasonic testing (UT) uses _____ sound waves to penetrate a metal object.
   a) Subsonic
   b) Sonic
   c) Supersonic
   d) Ultrasonic
   Ans: (d)

Interview Preparedness

Question 1. What's a 3D printing?
Response: 3D printing is the process of constructing a three-dimensional model by adding material automatically rather than removing material as in case of drilling or machining operatsons. This technique is also known as additive manufacturing, which was first implemented at the end of the 1980s and its first use for commercial operation was started as a quick prototyping tool in the automotive and aerospace industries. 3D Systems were developed by Charles Hull, who later co-founded, and invented a stereo lithography (SLA) method.3D Systems were launched in1988 as the first commercial 3D printer using SLA technology.

A number of industrial 3D printing associations were founded in the early 1990s and all of them were developed with new and innovative processes. Just three of the big 3D printing firms of the period when the main use for 3D printing was commercial are still on the market. The use of 3D printing was not widely available to the public until 2009. Several of the open source projects opened the door to affordable 3D desktop printers using Fused Deposition Modelling (FDM) technology. Then in the years following 2009, new and other companies began to invent, build and develop the 3D printer for the consumer/desktop to the point that it is still usable.

Question 2. How does 3D printing work?
Response: "Additive process" is used to create a 3D printed object. The succeeding layers are formed by laying down of material, until the three-dimensional structure is finished.

Question 3. What are the boundaries of 3D printing?
Response: While it is now widely applied in the enclosure of prototyping, 3D printing is only a few years away from making a breakthrough in the world of production engineering. This breakthrough would make it possible for 3D printing to expand from prototyping, with the exception of selected parts already produced by 3D printers, to be widely used in day-to-day manufacturing processes around the world.

The only factor that stops this from happening faster is the fairly long time it takes to 3D print something that can just as easily be manufactured using conventional methods.

Three-dimensional printing is limited to 3D printer dimensions as well. Although there are some pretty large 3D cement printing units, for example, high-quality and precision parts are limited to smaller machines, which can also be very expensive depending on what they are designed to be capable of.

Question 4. What is the difference between additive manufacturing and 3D printing?
Response: The short answer is no. The word '3D printing' is used to put either thickness of UV-curable photopolymer resin or the binding material on a powder layer in the process of a powder bed with inkjet printer heads. Nowadays, however, all additive manufacturing techniques are used universally.
"Additive manufacturing" refers to the more technical or correct way to automated building a digital file from scratch of a 3D object.

Question 5. What's the difference between 3D printing and 4D printing?
Response: 4D printing is a subset of 3D printing. In "normal" 3D printing, the final product is static unless few soft materials are applied and are meant to remain in that fashion. 4D printing is an approach to "program" the material to change form or performance when the correct impulse is provided.
For example, running slippers can be created to change the way they feel or fit depending on the activity of the wearer. Clothing could also change its functionality, depending on the weather. Although these products and solutions are not yet available in the market, various research institutions are working to improve these technologies.

Question 6. Who invented 3D printing?
Response: The co-founder of 3D Systems, Charles (Chuck) Hull, invented a solid imaging process which is today known as stereolithography (SLA). SLA was the first profit making 3D printing method. Charles first thought about the idea in 1983 while using UV light to amalgamate tabletop coatings. Patent was granted for this process in 1986. Charles Hull also invented the STL file configuration, which is the most common 3D file printing format.

Question 7. Are 3D printed goods as similar as good or manufactured products?
Response: 3D printing has made it possible to improve certain products beyond the capabilities of traditional manufacturing processes such as jet engine components. However, on a general level, it would be an abstract statement to suggest that a certain 3D printed product is better or worse than its traditional counterpart. From an economic point of view, 3D printing is on its way to becoming a very efficient, resourceful and cost-effective means of production and is likely to surpass conventional manufacturing processes in the near future.

Question 8: What are the possible applications of 3D printing in education?
Response:

  i. Yes, 3D printing has already been introduced to students at an early age.
  ii. 3D printing enables the physical manifestation of the thoughts and ideas of someone else. Hands-on experience makes a technical subject fun and attractive, even for people who are not (yet) particularly interested in design, architecture, computer science or engineering and many other fields

of study. To some extent, dry and theoretical and boring class, can be made interesting through an attractive and practical experience.

iii. China had introduced all its elementary schools with 3D printers in 2017. The initial engagement of students in newer technologies will lead this a reality.

iv. In Japan, Kabaku and Microsoft corporations also designed a 3D printing and programming learning device. In the popular Mine craft video game, students can develop framework in the game and output their creations to a 3D printer. This helps students realize the potential impact that their imaginations and ideas can make them realize that design is a trial-and-error process, often involving several iterations before it gets right.

v. There is no debate as to whether or not 3D printing can be applied in the classroom.

Question 9. Is Unsupported Filaments going to ruin 3d printer?

Response: The advice is to operate a 3D printer in accordance with the manufacturer's guidelines. Often, the use of the wrong filament can result in damage to the hot end of the 3D printer. Generally, ensure that the temperature of the nozzle is not increased to temperatures above the given specifications and that only the filaments recommended by the manufacturer are used. For example, if the nozzle is not built of material suitable for carbon fiber, the carbon fiber-infused filament will evaporate the nozzle. (Since the 3D printers are modular, a worn-out nozzle can be replaced.)

Question 10. What are the types of 3D printing filaments?

Response: There are different types of 3D printing filaments. First and foremost, they come in two conventional diameters: 1.75 mm and 3.00 mm.

PLA and ABS are the highest usual materials used in FDM 3D printing. These materials are attractive due to their ease of application (ABS being a little more complicated than 3D printing) and economy.

But as far as the materials considered for printing, there are almost no boundaries. It is achievable to print in natural metal, food and all types of thermoplastics where metal and food does not come in filament form for various reasons.

Question 11. What kind of software is needed for 3D printing?

Response: There is the need for at least a 3D slicer to prepare an existing 3D model for 3D printing. Most 3D printers come with their own cutting software. Cura, the most versatile Simplify3D, is the most popular 3D open source slicer.

For designing 3D models or modifying existing models 3D modelling software is needed.

Question 12. How to create 3D Printable Models?

Response: First, 3D models can be made using 3D modelling software or real-life objects, which are then scanned and transformed into 3D model files with specialized software generally provided by the scanning device manufacturer. After that, 3D models need to pass through a 3D slicing program to become a 3D printable model.

Question 13. What do you mean by 3D Printing Model?
Response: A 3D printing model is a file that can be read and interpreted by a 3D printer. It is used to inform a 3D printer where to place the nozzle to create a physical object. The model file consists of geometric information which will have to be understood by slicing software that turns the geometric input into the commands that the printer can operate.

Question 14. How to start when designing a 3D print model?
Response: Download and install a 3D modeling software. There are a lot of good, free 3D modeling programs out there that are perfect for beginners. Start with uncomplicated models, like a cube, and then through more intricate articles.

Question 15. What are the most usual 3D file formats for printing?
Response: There are very less 3D file formats available for printing. Some of the most common formats are:

   i. STL: This is the highest versatile 3D model format, supported by all cutting tools.
   ii. OBJ: The OBJ file configuration is also important – and supported by all major slicing tools. In contrast to STL, OBJ is able to store color and texture profiles; it's likely to become more popular when multicolor printing starts.
   iii. PLY: PLY, the Polygon file composition, was initially applied to store 3D scanned objects.
   iv. 3MF:3MF is a latest file format developed by Microsoft, Autodesk, HP and Shape Ways.

Question 16. How to change the current 3D printing model?
Response: Modify current 3D models by importing a file to a 3D modeling program. Make all the desired changes and save progress and export the 3D model.
There are also extraordinary tools for editing STL files which are not full-blown CAD programs – and are therefore easier to apply.

Question 17. What's the minimal wall thickness in case of a 3D print model?
Response: There are some important things to have in mind when deciding on the wall thickness of print.

   i. First of all, for an extremely detailed model of a building, for example, scaling down the model for 3D printing will scale down every aspect of the model, including tiny details and wall thicknesses. There is a point where the 3D printer is no longer capable of printing such fine details. This could result in super fragile prints and missing details. Therefore, while developing a file for printing, it is significant to know which parameters to change in order to obtain better results.
   ii. The later aspect to examine is the material and machinery that areused.
   iii. In FDM technology, the minimum wall thickness is typically between 0.05 mm and 0.1 mm – depending on the resolution supported of 3D printer.

This does not mean, however, that the walls printed at that thickness will be stable at all.

iv. The shape of 3D model has been a critical aspect as well. For illustration, if 3D model has an object with a more overhang, the structures shielding the overhang are required to be thick enough not to pervert or crash. Also, it is important to make sure that the shielding walls are capable enough to hold the weight above them. In general, 1 mm is the lower constraint for wall thickness, hence 2 mm is recommended.

Question 18. Why does a 3D printing model have to be watertight?

Response: "Water tightness" in case of 3D printing model refers to a 3D model mesh. The mesh itself must have no holes in it. Otherwise, a 3D printer may not be capable of printing a model. This kind of problem can be fixed by using a 3D model repair tool.

Question 19. Can a 2D photo be used to get a 3D printed model?

Response:

i. Yes, but the results will be different. Also, it really depends on the scanning method chosen. Many 3D scanning machines use a series of 2D photos to develop or create a substantive 3D model of the object.

ii. Also, most of the 3D-printed figures are prepared from a multitude of 2D pictures that are stitched together.

iii. "Scandi" is an app that allows users to take 360° perspective drawings. Images can then be printed in 3D color in the mode of a sphere. This allows for a greater mesmerizing photo viewing experience. Apps also permit the users to take 3D photos to design 3D models of people, animals or objects and if 3D model to be developed, Scandi will print it.

Question 20. How to reduce the cost of 3D printing?

Response: There are a number of ways to reduce 3D printing costs. Variables such as material, size, and print filling can make a huge price difference over time if they are optimized.

Question 21. What are the types of materials suited for 3D printing?

Response: Printing with various kinds of materials is possible. Most frequently known, 3D desktop printers can print with any number of thermoplastics and thermoplastics mixed amidst other materials like wood fibre, metal powder, dark compound glow and many more. Scientists are even experimenting with printing biomaterials which is an effort to ultimately print 3D human organs for transplantation.

Question 22. Which kind of materials can't be printed in 3D?

Response: It's difficult to think of any material that can't be printed in 3D. Theoretically, given the correct precedence, almost any type of material can be printed in 3D format. But practically liquids cannot be printed in 3D, unless the room in which 3D printing are used is minimum below 0° C and provided with 3D printer along with essential modifications. Other materials, such as natural gaseous materials, cannot also be printed in 3D under normal conditions.

Question 23. What kind of resolution can a 3D printer print?
Response: The 3D printer resolution is divided into two parts. First, the XY resolution is the minimal movement the nozzle that can be made on the X and Y axes. Later, there is vertical resolution to which the least coat thickness of the 3D printer can be set. This happens for 3D FDM printers.
Maximum resolution (quality) varies greatly built upon the 3D printing mechanism used.

Question 24. What's the successful way to print Fully Functional 3D Parts?
Response: The successful path to print fully functional 3D parts in excellent quality is to apply professional 3D printing services. In order to safeguard, the service can naturally produce design, and it would be a best idea to contact the customer service provider.

Question 25. What are the advantages of 3D printing compared to injection molding?
Response: A major advantage of 3D printing compared to injection molding is the cost advantage. Injection molding requires the first production or formation of a mold, which is a costly and delicate process. However, as the cost of injection molding decreases with increasing production volumes, there is usually a point where injection molding makes more sense.
Also, some 3D printable geometric shapes simply cannot be produced by injection molding.

Question 26. What are the differences between 3D printing and CNC milling?
Response: The prominent distinction amongst 3D printing and CNC milling is the starting point: with 3D printing, the manufacturing process starts from nothing, the objects are constructed by adding layers of materials. In CNC milling, the process begins with a block of material and is finished by removing the material until the desired shape has been formed.

Question 27. What is not suited for 3D printing?
Response:

    i. As far as 3D printing is considered, much is possible – but even more is impossible. An industrial-grade polished surface cannot be achieved by applying an FDM 3D printer. It has to be post-processed.

    ii. An industrial-grade polished surface cannot be achieved by using an FDM 3D printer. It has to be post-processed.

    iii. At a hypothetical level, 3D printing of an operational, intricate electro-mechanical machine or products like automobile vehicles is not yet possible at a single stretch, even though there is no problem with 3D printing of individual parts. Researchers across the globe are already working on the production of 3D printers.

iv. Also, bio-printing is not developed enough for 3D printing of fully functional, tailor-made human organs.

Question 28. Is 3D printed gold the same as "normal" gold?
Response: First thing is that, gold is not a printable 3D material. Still, 3D printing is progressively being used to make gold objects by 3D printing a wax model, which is then cast in solid gold. In other words, gold printed in 3D or not, will be as valuable as the current market price of gold is considered. In few cases, some famous jewelers' handmade jewelry are selling considerably higher than their weight in gold.

Question 29. Is 3D printing of food possible?
Response: Yes, it's possible to print 3D food and it's already done. The requirement to do and make food is perplexing. The food paste is then brimmed into a syringe-like container and passes over a nozzle onto a plate or platform.

There are, however, some drawbacks. Even great looking food can be created; it's almost impractical for some people to print a complete meal in 3D – printers are simply slower. Still, the chefs are embracing 3D printing technology quickly.

Question 30. Can chocolate be printed in 3D?
Response: Yes, there are possibilities for chocolate to be printed in 3D format and may even be better when compared with other foods. Since chocolate can melt, it can be passed through a heated extruder and cooled to a solid like thermoplastics. Unfortunately, there is no 3D chocolate printer available at present in the market, but using all-in-one printers chocolates can be designed.

Question 31. What's the prominent latest thing which can be done with 3D printing?
Response:

i. 3D printing has opened up many opportunities, especially in the field of aerospace and medicine. Certain jet engine components with complex internal geometry can now be manufactured using 3D printing. These parts would have been almost impossible to manufacture using conventional methods.

ii. Many space exploration projects/concepts depend heavily on 3D printing. For example, the Russian and European Space Agency is planning to build 3D bases on the moon using moon soil rather than sending all materials from the earth. This is going to result in saving lots of money.

iii. On the medical front, scientists are working on bio-printing, which will permit 3D printing of living tissues, bones and organs.

Question 32. What are the areas of 3D industrial printing?
Response: 3D printing is used in the aerospace, automotive, medical, architectural, consumer goods/electronics, defence and education sectors.

Interview Preparedness

Since there are many applications in which 3D printing can be done cost-effectively, 3D printing has found its place in several industries. Time saved by 3D printing an architectural model, rather than constructing or milling is tremendous.

Question 33. What's the biggest thing that was printed in 3D?

Response: The current holder of the Guinness World Record title "Largest 3-D printed object" is the Oak Ridge National Laboratory (ORNL). They possess a 3D printing tool to be applied in the production of the Boeing 777X, a latest and improved version of the popular and successful Boeing 777 aircraft. The solid object is 17.5 ft long, 5.5 ft wide and 1.5 ft high according to the definition of the record title.

Question 34. Is 3D printing going to change the world?

Response: An industry believes that 3D printing will essentially have a major impact on the manufacturing sector. Dispersed and regional manufacturing fused with mass customization will certainly reform the design and manufacture of products along with process logistics.

Question 35. How to print text or nameplates using 3D?

Response: This sounds pretty standard, but it could be quite difficult without having any experience using 3D modelling software. Many 3D modelling platforms have a text-based function, even in different fonts. It is essential to note that a suitable base for 3D printing is required. This is going to keep the letters in place and keep them from falling over.

References

3D Systems. Stereolithography and selective laser sintering machines. www.3dsystems.com
3D Systems Inc. Design guide. 2015. https://www.3dsystems.com/resources.
Talal Al-Samman, Material and process design for lightweight structures, *Metals* 2019, 9, 415. doi:10.3390/met9040415
Adedeji B. Badiru, Vhance V. Valencia, David Liu, *Additive Manufacturing Handbook Product Development for the Defense Industry*, 2017, CRC Press Taylor & Francis Group, Boca Raton, FL.
Cindy Bayley, Lennart Bochmann, Colin Hurlbut, Moneer Helu, David Dornfeld, Understanding error generation in fused deposition modeling. *Journal of Surface Topography: Metrology and properties*, 2015. doi:10.1088/2051-672X/3/1/014002
Olugbenga Solomon Bello, Kayode Adesina Adegoke, Rhoda Oyeladun Oyewole, Biomimetic materials in our world: a review, *IOSR Journal of Applied Chemistry*, Sep.–Oct. 2013, 5(3), 22–35. e-ISSN: 2278-5736
V. Birman, L. W. Byrd, Modelling and analysis of functionally graded materials and structure, *ASME Applied Mechanics Reviews* 2007, 60(5), 195–216.
Joran W. Booth, Jeffrey Alperovich, Pratik Chawla, Jiayan Ma, Tahira N. Reid, Karthik Ramani, The design for additive manufacturing worksheet, *Journal of Mechanical Design* October 2017, 139, 100904-1. Copyright VC 2017 by ASME.
Helena N. Chia, Benjamin M. Wu, Recent advances in 3D printing of biomaterials, *Journal of Biological Engineering* 2015, 9, 4. ISSN: 754-1611, doi:10.1186/s13036-015-0001-4
J. Choi, O. C. Kwon, W. Jo, H. J. Lee, M.-W. Moon, 3D printing and additive manufacturing, *International Journal of Precision Engineering and Manufacturing-Green Technology*, 2015, 2(4), 159–167.
C. K. Chua, K. F. Leong, *Rapid Prototyping: Principles and Applications in Manufacturing*. Wiley, New York, 1998.
Joaquim de Ciuranaa, Ldia Serenóa, Èlia Vallèsa, Selecting process parameters in RepRap additive manufacturing system for PLA scaffolds manufacture, *Procedia CIRP* 2013, 5, 152–157.
Arup Dey, Nita Yodo, A systematic survey of FDM process parameter optimization and their influence on part characteristics, *Journal of Manufacturing and Materials Processing* 2019, 3, 64. doi:10.3390/jmmp3030064, www.mdpi.com/journal/jmmp
Andreas Gebhardt, *Understanding Additive Manufacturing Rapid Prototyping, Rapid Tooling, Rapid Manufacturing*, Hanser Publishers, Munich; Hanser Publications, Cincinnati.
Ian Gibson, David Rosen, Brent Stucker, *Additive Manufacturing Technologies 3D Printing, Rapid Prototyping, and Direct Digital Manufacturing*, Springer, New York. ISBN 978-1-4939-2112-6, ISBN 978-1-4939-2113-3 (eBook). doi:10.1007/978-1-4939-2113-3.
Muhammad Harris, Johan Potgieter, Richard Archer, Khalid Mahmood Arif, In-process thermal treatment of polylactic acid in fused deposition modeling, *Materials and Manufacturing Processes* 2019, 34(6), 701–713.
P. F. Jacobs, *Rapid Prototyping & Manufacturing, Fundamentals of Stereolithography*. Society of Manufacturing Engineers, New York, 1992.
Udayabhanu Jammalamadaka, Karthik Tappa, Recent advances in biomaterials for 3D printing and tissue engineering, *Journal of Functional Biomaterials* 2018, 9, 22. doi:10.3390/jfb9010022

K. G. Jaya Christiyana, U. Chandrasekharb, K. Venkateswarluc, A study on the influence of process parameters on the mechanical properties of 3D printed ABS composite, *IOP Conference Series: Materials Science and Engineering* 2016, 114, 012109. doi:10.1088/1757-899X/114/1/012109

W. Jo, D. H. Kim, J. S. Lee, H. J. Lee, M.-W. Moon, *RSC Advances* 2014, 4, 31764–31770.

J. P. Kruth, P. Mercelis, J. Van Vaerenbergh, Binding mechanisms in selective laser sintering and selective laser melting, *Rapid Prototyping Journal* 2005, 11(1), 26–36.

John J. Laureto, Joshua M. Pearce, Open SourceMulti-Head 3D printer for polymer-metal composite component manufacturing, *Technologies* 2017, 5, 36. doi:10.3390/technologies5020036

W. Liu, C. Wu, W. Liu, W. Zhai, J. Chang, The effect of plaster ($CaSO_4 \cdot 1/2H_2O$) on the compressive strength, self-setting property, and in vitro bioactivity of silicate-based bone cement, *Journal of Biomedical Materials Research B: Applied Biomaterials* 2013, 101B, 279–286.

Céline A. Mandon, Loïc J. Blum, Christophe A. Marquette, 3D–4D printed objects: new bioactive material opportunities, *Micromachines* 2017, 8, 102; doi:10.3390/mi8040102, www.mdpi.com/journal/micromachines

Y. Mao, K. Yu, M. S. Isakov, J. Wu, M. L. Dunn, H. Jerry Qi, Sequential self-folding structures by 3D printed digital shape memory polymers, *Scientific Reports* 2015, 5, 13616.

John O. Milewski, *Additive Manufacturing of Metals From Fundamental Technology to Rocket Nozzles, Medical Implants, and Custom Jewelry*, ISSN 0933–033X ISSN 2196–2812 (electronic) Springer Series in Materials Science ISBN 978-3-319-58204-7 ISBN 978-3-319-58205-4 (eBook). doi:10.1007/978-3-319-58205-4

Thabiso Peter Mpofu, Cephas Mawere, Macdonald Mukosera, The impact and application of 3D printing technology, *International Journal of Scientific and Engineering Research (IJSR)* 2012, 3, 358. ISSN (Online): 2319–7064 Impact Factor.

Thabiso Peter Mpofu, Cephas Mawere, Macdonald Mukosera, The impact and application of 3D printing technology, June 2014, 3(6). Paper ID: 02014675.

E. J. Murphy, R. E. Ansel, J. J. Krajewski, Investment casting utilizing patterns produced by stereolithography, DeSoto, Inc, US Patent 4,844,144, 4 July 1989.

R. Rabiei, A. K. Dastjerdi, M. Mirkhalaf, F. Barthelat, Hierarchical structure, mechanical properties and fabrication of biomimetic biomaterials. *Biomimetic Biomaterials, Structure and Applications*, September 2013. doi:10.1533/9780857098887.1.67

D. Raviv, W. Zhao, C. McKnelly, A. Papadopoulou, A. Kadambi, B. Shi, S. Hirsch, D. Dikovsky, M. Zyracki, C. Olguin, R. Raskar, S. Tibbits, Active printed materials for complex self-evolving deformations, *Scientific Reports* 2014, 4, 7422.

P. Rochusa, J.-Y. Plesseriaa, M. Van Elsenb, J.-P. Kruthb, R. Carrusc, T. Dormalc, Newapplications of rapid prototyping and rapid manufacturing (RP/RM) technologies for space instrumentation, *International Journal of Acta Astronautica ScienceDirect* 2007, 61, 352–359.

Raju B. Sa, U. Chandra Shekarb, K. Venkateswarluc, D. N. Drakashayanid, Establishment of Process model for rapid prototyping technique (Stereolithography) to enhance the part quality by Taguchi method, *Procedia Technology*, 2014, 14, 380–389.

Swapnil Sayan Saha, Ariful Islamy, Shape memory material- concepts, recent trends and future directions, 2018 Joint 7th International Conference on Informatics, Electronics & Vision (ICIEV).

Anoop Kumar Sood, R. K. Ohdar, S. S. Mahapatra, Improving dimensional accuracy of Fused Deposition Modelling processed part using grey Taguchi method, *Materials and Design* 2009, 30, 4243–4252.

Rupinder Singh, Ranvijay Kumar, Ilenia Farina, Francesco Colangelo, Luciano Feo, Fernando Fraternali, Multi-material additive manufacturing of sustainable innovative materials and structures, *Polymers* 2019, 11, 62. doi:10.3390/polym11010062, www.mdpi.com/journal/polymers

Rupinder Singh, Sunpreet Singh, Fused deposition modelling based rapid patterns for investment casting applications: a review, *Rapid Prototyping Journal* 2016. doi:10.1108/RPJ-02-2014-0017

T. S. Srivatsan, T. S. Sudarshan, *Additive Manufacturing Innovations, Advances, and Applications*, CRC Press Taylor & Francis Group, Boca Raton, FL.

Stratasys. Fused deposition modelling. www.stratasys.com

A. Sydney Gladman, E. A. Matsumoto, R. G. Nuzzo, L. Mahadevan, J. A. Lewis, *Nature Materials* 2016, advance online publication. S. Tibbits *Archit. Des.* 2014, 84, 116–121.

S. Tamas-Williams, I. Todd, Design for additive manufacturing with site-specific properties in metals and alloys, *Scripta Materialia* 2016, 135, 105–110.

A. E. Tontowi, L. Ramdani, R. V. Erdizon, D. K. Baroroh, Optimization of 3D-printer process parameters for improving quality of polylactic acid printed part, *International Journal of Engineering and Technology*, Apr–May 2017, 9(2). doi:10.21817/ijet/2017/v9i2/170902044

C. H. Venu Madhav, R. Sri Nidhi Hrushi Kesav, Y. Shivraj Narayan, Importance and utilization of 3D printing in various applications, 2nd National Conference On Developments, Advances & Trends in Engineering Science [*NC- DATES 2K16*].

C. H. Venu Madhav, R. Sri Nidhi Hrushi Kesav, Y. Shivraj Narayan, Importance and utilization of 3D printing in various applications, *International Journal of Modern Engineering Research* 2016, ISSN: 2249–6645.

Juho-Pekka Virtanen, Hannu Hyyppä, Matti Kurkela, Matti Vaaja, Petteri Alho, Juha Hyyppä, Rapid prototyping—a tool for presenting 3-dimensional digital models produced by terrestrial laser scanning, *ISPRS International Journal of Geo-Information* 2014, 3, 871–890. doi:10.3390/ijgi3030871

Che Chung Wang, Ta-Wei Lin, Shr-Shiung Hu, Optimizing the rapid prototyping process by integrating the taguchi method with the grey relational analysis, *Rapid prototyping Journal* 2007, 13/5, 304–315.

David Ian Wimpenny, Pulak M. Pandey, L. Jyothish Kumar, *Advances in 3D Printing & Additive Manufacturing, Technologies* 2017, Springer, New York. ISBN 978-981-10-0811-5, ISBN 978-981-10-0812-2 (eBook). doi:10.1007/978-981-10-0812-2

Wenzheng Wu, Wenli Ye, Peng Geng, Yulei Wang, Guiwei Li, Xue Hu, Ji Zhao, 3D printing of thermoplastic PI and interlayer bonding evaluation, *Materials Letters* 2018, 229, 206–209.

Li Yang, Keng Hsu, Brian Baughman, Donald Godfrey, Francisco Medina, Mamballykalathil Menon, Soeren Wiener, *Additive Manufacturing of Metals: The Technology, Materials, Design and Production* 2017, ISSN 1860–5168 ISSN 2196-1735 (electronic) Springer Series in Advanced Manufacturing ISBN 978-3-319-55127-2 ISBN 978-3-319-55128-9 (eBook), doi:10.1007/978-3-319-55128-9

Muhammad Jamshaid Zafar, Dongbin Zhu, Zhengyan Zhang, 3D printing of bioceramics for bone tissue engineering, *Materials* 2019, 12, 3361. doi:10.3390/ma12203361, www.mdpi.com/journal/materials

Index

3-Dimensional, 1
3-Dimensional printing, 2
3D printing technology, 1
3D systems' ColorJet printing (CJP)
        technology, 92
3D systems' MultiJet printing system (MJP), 62
3D systems' selective laser sintering (SLS), 89
3D systems' stereolithography apparatus
        (SLA), 55

Acrylonitrile butadiene styrene, 59, 109
Adaptive raster milling, 228
Additive manufacturing, 1
Advanced AM materials, 151
Aerospace and defense, 169
Aerospace industry, 169
Arcam's electron beam melting (EBM), 95
Arts and architecture, 182
Asthetics, 108
Automotive industry, 171

Bead blasting, 164
Bending, 208
Binder jetting, 12, 13, 200
Binder jetting methods, 115
Bio-ceramics, 138
Biomaterials, 129
Biomedical, 130
Biomimetic material, 135
Biomimetics, 133
Bond mechanisms, 204
Building information modeling, 181
Build material supports, 153
Bulk deposit defects, 247

CAD model verification, 167
Carbon fiber reinforced plastic, 129
Carbon fiber reinforced thermoplastic, 129
Ceramics, 111, 137
CMET's solid object ultraviolet-laser printer
        (SOUP), 65
Coin industry, 178
Cold hibernated elastic memory, 143
ColorJet printing, 18
Color models, 45
Composite materials, 127
Compound annual growth rate, 4
Computer-aided design, 25
Computer tomography, 129
Concept laser's LaserCUSING, 97

Consolidation of the part, 171
Construction, 183
Continuous liquid interphase production, 16
Coordinate measurement machines, 256
Cubic technologies' laminated object
        manufacturing (LOM), 80
Customization, 172

Destructive test, 248
Destructive test methods, 258
Development of additive manufacturing, 1
Digital light processing, 6, 8
Digital micromirror device, 8
Directed energy deposition, 15, 200
Direct hard tooling, 169
Direct metal deposition, 8
Direct metal laser melting, 97
Direct metal laser sintering (DMLS) materials, 121
Direct printing, 176
Direct slicing of the CAD model, 44
Direct soft tooling, 168
Do-it-yourself, 146
Drop on demand, 6, 12

Eddy-current testing, 258
Edison welding institute, 84
Electrode discharge machining, 244
Electron beam free-form fabrication, 16
Electron beam melting, 18
Electro optical systems, 93
Elephant's foot, 210
EnvisionTEC's Bioplotter, 66
EnvisionTEC's Perfactory, 63
EOSINT M 280 system, 93
EOS's EOSINT system, 93
Epoxy coating, 166
Epoxy infiltration, 166
Esthetic improvements, 232
Extrusion-based printing methods, 116

Fiber reinforced plastic, 128
Finite element analysis, 47
Flow analysis, 168
Food, 187
Form and fit, 168
Format specifications, 28
Functionally graded materials, 114, 115
Fused deposition modeling, 6
Fused deposition modeling from Stratasys, 60
Fused filament fabrication, 109

Generic composite materials, 128
Geographic information system, 180
Glassfaserverstärkter Kunststoff, 128
Glass fiber reinforced plastic, 128
Glass reinforced plastic, 128
Governing bonding mechanism, 201

Heat treatments, 164
Hewlett-Packard, 5
High impact polystyrene, 112
History of biomimetic materials, 135
Hole drilling, 232
Hot isostatic pressing, 239

Indentation-polishing-heating, 144
Indirect hard tooling, 169
Indirect soft tooling, 168
Inkjet printing, 118
Investment casting, 173

Jewelry industry, 172

Laminated object manufacturing, 138
Laser engineered net shaping, 95, 121
Laser metal deposition, 116
Low-temperature deposition, 116
Low-volume production, 170

Machining strategy, 228
Magnetic resonance imaging, 129
Mask projection vat photopolymerization, 63
Material efficiency, 170
Material extrusion, 6, 200
Material jetting, 11, 12, 201
Material jetting machines, 61
Material jetting printing methods, 118
Medium-density fiberboard, 128
Messy first layer, 209
Metal AM processes and materials, 119
Metal binder jetting, 14
Metals, 114, 161
Micro-electromechanical systems, 141
Multijet printing, 17
Multiple materials, 45, 114
Musical instruments, 186

Nano-electromechanical systems, 141
Nondestructive test, 258
Nondestructive test methods, 255
Nylon, 108

Occupational Health and Safety Organization, 262
Optomec's laser engineered net shaping
        (LENS), 95
Out of filament, 205
Over-extrusion, 206
Overview of 4D printing, 264

Painting, 167
Penetrant testing, 252
PET/PETG, 112
Plating, 163
Polyamides, 123
Polyaryletherketons, 90
Polycarbonate, 75
Polyethylene terephthalate, 112
Polylactic acid (PLA), 59, 110
Polymer, 159
Polyphenylsulfone, 166
Polystyrenes, 90
Polyvinylidene Fluoride, 117
Post-processing and finishing, 244
Powder bed fusion, 6, 8, 199
Powder bed fusion (metals), 10
Powder bed fusion (polymers), 8
Powder-bed systems, 120
Powder-fed systems, 120
Powder production, 91
Powder reclamation, 243
Pre-production parts, 168
Process design, 213
Proof of concept, 167

Radio frequency identification, 4
Radiographic testing, 257
Rapid freeze prototyping, 69
Rapid prototyping, 55
RegenHU's 3D bioprinting, 67
Resin, 109
Room temperature vulcanization, 233

Sand binder jetting, 13
Sanding, 163
Scaling, 168
Selective deposition lamination, 16
Selective heat sintering, 16
Selective laser melting, 16
Selective laser sintering, 16
Shape memory alloys, 141
Shape memory composites, 146
Shape memory hybrids, 146
Shape-memory materials, 139
Shape memory polymers, 142, 146
Sheet lamination, 14, 200
Shot peening, 164
Sintering mechanisms, 91
Snapped filament, 211
Solid freeform manufacturing, 69
Solid object ultraviolet-laser printer, 17
Solidscape's BenchTop system, 75
Solvent dipping, 166
Stainless steel, 111
Stereolithography, 17
Stereolithography (SLA), 8
Stereolithography (SLA) apparatus, 16

Stereolithography (SLA) materials, 124
Stereolithography (SLA) methods, 115
STL ASCII format, 29
STL binary format, 30
STL file format, binary/ASCII, 26
STL file manipulation, 41
Stratasys fused deposition modeling (FDM), 75
Stratasys PolyJet, 59
Support material removal, 221
Support materials, 152
Survey of software functions, 46
Synthetic support removal, 25

Tableware industry, 180
Technology improvement, 3
Temperature memory effect, 142
Thermoplastic elastomers, 90
Thermoplastics, 113
Thermosets (resins), 114
Titanium, 111

Titanium aluminum, 4
Tumbling, 165

Ultra-high-molecular-weight polyethylene, 128
Ultrasonic additive manufacturing, 201
Ultrasonic consolidation, 83
Ultrasonic testing, 258
Ultraviolet, 12

Vapor smoothing, 166
Vat photopolymerization, 7, 201
Vat polymerization, 6
Vibratory systems, 165
Visualizing objects, 167

Warping, 208
Weapons, 186
Weight reduction, 170

Zygomatic implant, 70

Printed in the United States
By Bookmasters